Proceedings of the
Steklov Institute of Mathematics
Volume 199 Issue 1 of 6 1995

Relaxation Oscillations in Mathematical Models of Ecology

By A. Yu. Kolesov and Yu. S. Kolesov

Edited by E. F. Mishchenko

A Translation of

ТРУДЫ
МАТЕМАТИЧЕСКОГО ИНСТИТУТА
имени В. А. СТЕКЛОВА

ТОМ 199 1993

ABSTRACT: This book presents for the first time a systematic exposition of techniques for constructing relaxation oscillations and methods for investigating stability properties of certain classes of systems with delay. Distinctive features with no analogues in relaxation systems of ordinary differential equations are pointed out. The exposition is on the level of an analysis of significant examples from biophysics, mathematical ecology, and quantum physics. Many unsolved problems are posed. The contents should be useful for researchers and specialists interested in the theory and applications of relaxation oscillations.

1991 *Mathematics Subject Classification*. Primary 34–02, 92–02, 34K15, 34K20, 34K25, 34C05, 34C25, 92D25, 92D40; Secondary 34D45, 34C10, 34D05, 35Q80.

РОССИЙСКАЯ АКАДЕМИЯ НАУК

ТРУДЫ

МАТЕМАТИЧЕСКОГО ИНСТИТУТА
имени В. А. СТЕКЛОВА

CXCIX

А. Ю. Колесов, Ю. С. Колесов

РЕЛАКСАЦИОННЫЕ КОЛЕБАНИЯ В МАТЕМАТИЧЕСКИХ МОДЕЛЯХ ЭКОЛОГИИ

Под редакцией
академика Е. Ф. МИЩЕНКО

Ответственный редактор Трудов МИРАН
академик Е. Ф. МИЩЕНКО
Заместитель ответственного редактора Трудов МИРАН
доктор физико-математических наук Е. А. ВОЛКОВ

Рецензенты
доктора физико-математических наук
М. И. ЗЕЛИКИН, Н. Х. РОЗОВ

Москва «Наука» 1993

Correspondence between *Trudy Mat. Inst. Steklov* and *Proc. Steklov Inst. Math.*

Russian imprint	Vol.	English imprint	Russian imprint	Vol.	English Imprint	Issue	Russian imprint	Vol.	English Imprint	Issue	Russian imprint	Vol.	English Imprint	Issue
1966	74	1967	1971	109	1974		1979	144	1980	2	1988	179	1989	2
1965	75	1967	1970	110	1972		1980	145	1981	1	1987	180	1989	3
1965	76	1967	1970	111	1972		1978	146	1980	3	1988	181	1989	4
1965	77	1967	1971	112	1973		1980	147	1981	2	1988	182	1990	1
1965	78	1967	1970	113	1972		1978	148	1980	4	1990	183	1991	4
1965	79	1966	1970	114	1974		1978	149	1981	3	1990	184	1991	2
1965	80	1968	1971	115	1974		1979	150	1981	4	1988	185	1990	2
1966	81	1968	1971	116	1973		1980	151	1982	2	1989	186	1991	1
1966	82	1967	1972	117	1974		1980	152	1982	3	1989	187	1990	3
1965	83	1967	1972	118	1976		1981	153	1982	4	1990	188	1991	3
1965	84	1968	1973	119	1976		1983	154	1984	4	1989	189	1990	4
1966	85	1967	1974	120	1976		1981	155	1983	1	1989	190	1992	1
1965	86	1967	1972	121	1974		1980	156	1983	2	1989	191	1992	2
1966	87	1967	1973	122	1975		1981	157	1983	3	1990	192	1992	3
1967	88	1969	1973	123	1975		1981	158	1983	4	1992	193	1993	3
1967	89	1968	1976	124	1978		1983	159	1984	2	1992	194	1993	4
1967	90	1969	1973	125	1975		1982	160	1984	1	1994	195	1995	2
1967	91	1969	1973	126	1975		1983	161	1984	3	1991	196	1992	4
1966	92	1968	1975	127	1977		1983	162	1985	1	1991	197	1993	1
1967	93	1970	1972	128	1974		1984	163	1985	4	1992	198	1994	1
1968	94	1969	1973	129	1976		1983	164	1985	2	1993	199	1995	1
1968	95	1971	1978	130	1979	4	1984	165	1985	3	1991	200	1993	2
1968	96	1970	1974	131	1975		1984	166	1986	1	1992	201	1994	2
1968	97	1969	1973	132	1975		1985	167	1986	2	1993	202	1994	4
1968	98	1971	1973	133	1977		1984	168	1986	3	1993	204	1994	3
1967	99	1968	1975	134	1977		1985	169	1986	4				
1971	100	1974	1975	135	1978		1984	170	1987	1				
1972	101	1975	1975	136	1978		1985	171	1987	2				
1967	102	1970	1976	137	1978		1985	172	1987	3				
1968	103	1970	1975	138	1977		1986	173	1987	4				
1968	104	1971	1980	139	1982	1	1986	174	1988	1				
1969	105	1971	1976	140	1979	1	1986	175	1988	2				
1969	106	1972	1976	141	1979	2	1987	176	1988	3				
1971	107	1976	1976	142	1979	3	1986	177	1988	4				
1968	108	1971	1977	143	1980	1	1986	178	1989	1				

This nonjournal serial does not always appear in strict sequential volume order in the original. Accordingly, the translation will occasionally reflect any gaps resulting thereby, as happens for example with the English imprints of 1984, 1985, and 1990.

CONTENTS

RUSSIAN CONTENTS*

*The American Mathematical Society scheme for transliteration of Cyrillic may be found at the end of index issues of *Mathematical Reviews.*

Preface

Two aspects of the problem should be distinguished in the theory of relaxation oscillations. The first, the content of the C-theory, has to do with determining a limit object to which the relaxation system converges, and with an accompanying formulation of conditions for the existence of a relaxation cycle. The other part of the theory is more complicated: its goal is to justify convergence to the limit object in smooth metrics. In particular, the problems of uniqueness and stability of a relaxation cycle are solved at the same time.

This scheme is realized in full scope for relaxation systems in a finite-dimensional space. However, it must be modified in the study of systems with delay for which the phase space is infinite-dimensional. We mention three typical cases. First, the simplest situation is realized when from a sufficiently large domain of initial conditions the solutions wind around a relaxation cycle whose multipliers, except for 1, converge to zero as the dynamic parameter decreases or increases. Here the C-theorem reduces to a proof of the existence of a cycle, while the C^1-theorem reduces to a proof of the behavior of the multipliers. The second situation is to a certain degree analogous to the classical situation: the relaxation system collapses in the uniform topology to the limit object, and an analysis of the variational equations deepens this connection. Finally, the most interesting and complicated case is possible when the relaxation system converges to the limit object only in the weak topology. In an important example from mathematical ecology it is shown that in this case countably many multipliers of the variational equation have nonzero limits.

We have limited our exposition mainly to a study of substantive examples. We leave to the beginning researchers various generalizations, extensions to larger classes of equations, and so on.

Introduction

The theory of relaxation oscillations in multidimensional systems of ordinary differential equations, our sole interest here, has its origins in the note [59] of Pontryagin and Mishchenko, which served as a basis for their subsequent publications [58] and [60]–[63]. This theory received its finishing touches in [28], [29], and [31]. We note especially [28], where it was shown that under a certain general position condition an arbitrary relaxation system of ordinary differential equations is a small perturbation in the C^1-metric of a relay system corresponding to it in a certain way. That paper also clarified the content of the preceding references: they deal with C-convergence of a relaxation system to a suitable relay system.

An arbitrary relaxation system of ordinary differential equations is thus a smooth perturbation of a relay system in the general case. On considering the question of relaxation oscillations in systems with delay we encounter a more diverse situation. In particular, no unified theory has been constructed for this class of evolution equations. In this connection the exposition to follow is on the level of substantive examples of independent and significant interest, which enable us to reveal important patterns.

This monograph, which consists of three chapters, summarizes the investigations of the authors over the past several years ([13], [14], [22], [25], [26], [35], [43]–[45]). The first chapter is devoted to a study of relaxation oscillations whose amplitude grows unboundedly as a small parameter decreases or a large parameter increases. Our first example is the classical equation of Hutchinson describing the variation in density of a species whose relaxation cycle is connected with the fertility of the population. A distinctive feature of a Hutchinson relaxation cycle is that, for the variational equation, all the multipliers except 1 approach zero exponentially as the Malthusian coefficient of linear growth increases. However, an intermediate case encountered in the study of a bilocal model of the Hutchinson equation with diffusion is also possible. The corresponding system of differential-difference equations has a two-dimensional torus on which the solutions wind around exponentially with a large exponent. On the torus itself, however, the motions are slow, which makes a complete analysis of the question more difficult. Finally, an even more complicated example is supplied by a mathematical model of the Belousov reaction. In it the characteristic exponents of the variational system are finite, and smallness of the multipliers is ensured by an asymptotically large period of oscillations.

Of greatest significance, however, is the study in Chapter I of a mathematical model of the problem of interaction of a predator witha strongly fertile prey. This example is standard: a class of initial conditions is singled out such that the solutions of the corresponding system of differential-difference equations are close in the C^1-metric to some two-dimensional relay system (in terms of the first-return operator, to some one-dimensional mapping).

In Chapter II we use the example of a mathematical model of the dynamics of insect populations to begin an investigation of properties of relaxation cycles of finite amplitude (that is, bounded above by a constant independent of the small parameter). The main feature of the differential-difference equation considered is convergence to a limit object in the weak topology, which makes it more difficult even to prove an existence theorem for cycles. A less trivial consequence is that countably many multipliers of the variational equation have nonzero limits as the small parameter tends to zero. At the end of the chapter we show that the equation under investigation has also a large supply of strongly unstable relaxation cycles. The latter means that the variational equations have multipliers with moduli increasing without limit as the small parameter decreases.

Chapter II should be regarded as central: the differential-difference equation investigated in it is of interest from many points of view. To the preceding paragraph we add, for example, that the problem of finding certain solutions of special equations arises in constructing the asymptotics of cycles, and this is typical for complex problems in the theory of relaxation oscillations.

In Chapter III we continue the study of relaxation cycles of finite amplitude. Here we investigate a system with delay that is of interest both in quantum physics and mathematical ecology. We emphasize that for a slowly oscillating cycle of the system the C-convergence theorem is analogous to the corresponding assertion in the finite-dimensional case. This circumstance explains the C^1-theorem, which asserts that only two multipliers of the variational system, one of them 1, have nonzero limits as the small parameter decreases without limit. A special device is used to prove also a theorem on the existence of a large number of rapidly oscillating relaxation cycles called cycles with short periods because of their specific nature. The question of their stability is also solved with the help of a specially developed technique. It is essential that they can be either stable or unstable. A more thorough analysis of the stability properties of the simplest of them showed that loss of stability takes place in an oscillatory way. In this connection we formulate the problem of constructing stable two-frequency oscillations. The solution of this problem involves a certain four-dimensional mapping.

At the end of Chapter III it is shown that the results bear a general character—the technique is applied to the investigation of relaxation cycles of differential-difference equations of rather general form.

The authors saw a need for an investigation on the monograph level of the theory of relaxation oscillations in ecological equations not only in connection with the importance of the problem, but also with a view to directing the attention of experts to a new and promising direction: the study of relaxation cycles of differential-difference equations. Therefore, it seems both timely and sufficiently important to give a detailed exposition of methods of analysis already worked out, in particular, a clarification of the connection with the classical case of ordinary equations.

Chapter I was written by Yu. S. Kolesov. Chapters II and III were written jointly. The results of the last two sections are due to A. Yu. Kolesov.

CHAPTER I

Relaxation Cycles of Asymptotically Large Amplitude

1. Some facts from mathematical ecology

1.1. The equations of a single-species biocenosis. The first mathematical model for describing the dynamics of the number of a species was proposed in 1798 by Thomas Malthus, who set forth the idea that under favorable conditions any species increases its number $N(t)$ according to an exponential law; that is,

$$\dot{N} = rN, \tag{1.1}$$

where the parameter r came to be called the Malthusian coefficient of linear growth. It should be noted, however, that as far back as 1202 Fibonacci proposed an analogous mechanism for population growth in the solution of a concrete ecological problem which involved the Fibonacci numbers. The Malthusian law has been well corroborated by experimental data in cases when the number of the population is not large, that is, as long as its size is not limited by accessible food and territory. In order to take inhibiting factors into account, L. A. Quetelet and P. F. Verhulst proposed in 1835 the equation

$$\dot{N} = r(1 - N/K)N, \tag{1.2}$$

which is now called the logistic equation. This equation contains the important parameter K (called the capacity of the habitat), which characterizes the average size of the population and depends mainly on the amount of food and the area of the habitat. The logistic law was rediscovered in 1920 by R. Pearl and L. J. Reed.

The law given by equation (1.2) works well in describing the dynamics of growth of microorganism populations. However, if we try to apply it to the dynamics of the number of abundant species of mammals, then we are clearly going beyond the limits of applicability. The fact of the matter is that the density (that is, the quantity N/K) of many species of mammals living, say, in Canada or Yakutia can vary sharply in the course of time ([66], [79]). To preserve the logistic law also for these species, Hutchinson [75] proposed in 1948 the equation

$$\dot{N} = r[1 - N(t-h)]N, \tag{1.3}$$

where, in contrast to (1.2), N is the density of the population at the current moment of time. In (1.3) the positive parameter h is the average age of sexually mature females in favorable years, and

$$r = \ln(1 + mp), \tag{1.4}$$

where m is the fraction of sexually mature females, and p is the average size of a litter [33]. We emphasize that equation (1.3) models the simplest situation when the

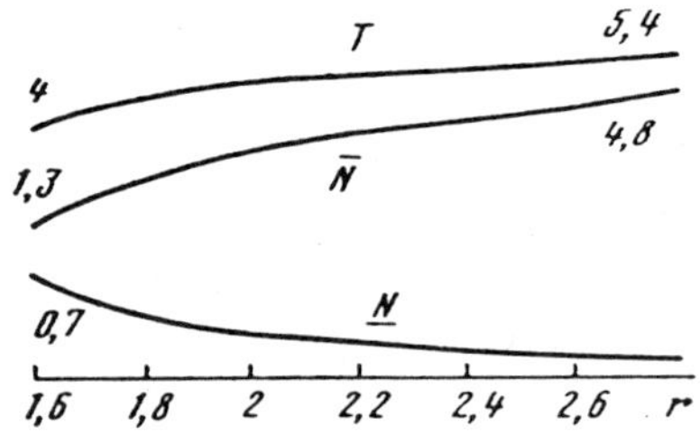

FIGURE 1.1

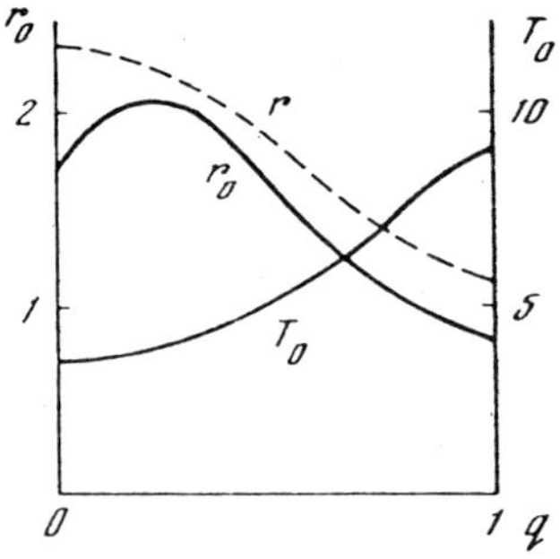

FIGURE 1.2

habitat is homogeneous, migration factors are not taken into account (more precisely, they are so significant that complete intermixing takes place), and the amount of accessible food is regularly replenished up to some level. It is not hard to see that undamped auto-oscillations appear in equation (1.3) under the conditions described as the product rh increases. The first results on the way to a rigorous proof of the existence of a cycle for $rh > \pi/2$ were obtained in 1961 by Jones [76], who made essential use of Wright's paper [80]. However, the question was completely cleared up only much later (see [73], where there is a bibliography). We remark also that equation (1.3) was related to the question of the density of the prime numbers in [74] and [81].

Figure 1.1 (where the unit of time is h) shows the dependence of the basic characteristics of a Hutchinson cycle on the Malthusian coefficient of linear growth.

The right-hand side of equation (1.3) is the difference between two terms, the first of which is connected with the birthrate factor according to the formula (1.4). The second term, which ensures a delayed internal feedback, is connected with limitations of food resources and territory for living. It thereby becomes clear that taking the more subtle age structure of the population into account leads to a differential-difference equation [38]

$$\dot{N} = r\left[1 - \sum_{j=1}^{n} \alpha_j N(t - h - j + 1)\right] N, \tag{1.5}$$

where the positive parameters $\alpha_j < 1$ are connected by the equalities

$$\alpha_j = q^{j-1}\alpha_1, \quad \alpha_1(1 + q + \cdots + q^{n-1}) = 1. \tag{1.6}$$

In [38] equation (1.5) was used for explaining certain features of the oscillations in the density of the white hare and the muskrat in Yakutia.

Figure 1.2 shows the dependence of the bifurcation value r_0 and the quantity $T_0 = 2\pi/\omega_0$ (where ω_0 is the excitation frequency of the auto-oscillations) on q, $0 < q < 1$, for $n = 4$ and $h = 0.9$. The dashed curve indicates the values of the Malthusian coefficient of linear growth for which the amplitude of the oscillations is equal to three (the graph is taken from [68], where the question of oscillations in the density of the Yakut squirrel and of its main food competitors is considered). It is quite clear that the oscillations are suppressed for relatively small q. The situation is analogous in the case of the muskrat population [38]. However, in the case of the white hare there is an increase in the size of the oscillations for any q due to long-lived individuals.

We proceed to the question of how to determine oscillations of individual age groups. Denote by $u(t, s)$ the density of individuals of age s at time t. It is natural to assume that this function satisfies the system of equations [37]

$$\frac{\partial u}{\partial t} + \varphi(t)u + \frac{\partial u}{\partial s} = 0, \qquad u(t, 0) = \psi(t)N(t), \tag{1.7}$$

where φ and ψ are the mortality function and the birthrate function, respectively. Obviously, the system (1.7) is supplemented by the relations

$$N(t) = \int_0^\infty u(t, s)\, ds, \qquad \lim_{s\to\infty} u(t, s) = 0. \tag{1.8}$$

From (1.7) we have

$$u(t, s) = N(t-s)\psi(t-s) \exp \int_{t-s}^{t} \varphi(\tau)\, d\tau. \tag{1.9}$$

Substituting (1.9) in the first equality in (1.8), we get the connection

$$\dot{N} = (\psi - \varphi)N \tag{1.10}$$

among φ, ψ, and N.

Thus, according to (1.9) and (1.10), in solving the problem of oscillations in the density of age groups we must know ψ or φ. It follows from biological considerations that the function

$$\varphi(t) = \Psi(\dot{N}/N) > \dot{N}/N$$

must be close to a constant for a low population density when the birthrate is high and the mortality low, and it must decrease strongly for a high density when, on the contrary, the birthrate is low and the mortality high. As is obvious, this property of the function Ψ must appear most noticeably when the Malthusian coefficient of linear growth increases. Numerical computations show that there is good agreement with the biological meaning if we take

$$\Psi(x) = \frac{1}{a} \exp[a(x - x_0)], \qquad a, x_0 > 0,$$

where $ax_0 < 1$ if $ar > 1$, and $\exp[a(r - x_0)] \geq ar$ if $ar \leq 1$.

The approach here to determination of the oscillations in the density of the age groups was tested, in particular, on the equation

$$\dot{N} = r\left[1 - \sum_{j=1}^{n} \alpha_j N(t - h - j + 1)\right] N - u(t)N, \tag{1.11}$$

where $u(t)$ is a certain nonnegative periodic function introduced in [36] for taking into account the influence of hunting on the dynamics of the variation in the density of the population. It turned out that there is a rejuvenation of the population irrespective of the choice of hunting strategy, and this agrees well with reality.

Finally, taking into account the migration factor leads to the boundary value problem

$$\frac{\partial N}{\partial t} = D\Delta N + r(1 - N_{t-h})N, \qquad \left.\frac{\partial N}{\partial \nu}\right|_{\partial\Omega} = 0, \tag{1.12}$$

where $N(t, x)$, $x = (x_1, x_2)$, is the population density at time t at a point x of a planar domain Ω with sufficiently smooth boundary $\partial\Omega$, Δ is the Laplace operator, $N_{t-h} = N(t-h, x)$, and ν is the normal direction at points of $\partial\Omega$. In (1.12) the diffusion coefficient D is proportional to the normal living area needed on the average by a separate individual [49]. From this it is clear that a desynchronization of the auto-oscillations necessarily arises when D is decreased; that is, a spatially heterogeneous structure appears in the homogeneous environment.

Both equation (1.3) and, especially, the boundary value problem (1.12) are rather complicated. However, the basic properties of (1.3) can be understood by considering the simplified model

$$N(t+1) = N(t)F(N), \qquad t = 0, 1, \dots,$$

where

$$F(N) = \exp\left[\lambda\left(1 - \frac{N(t) + N(t-1)}{2}\right)\right].$$

Similarly, in the simplest variant $\Omega = (0, 1)$ a good idea of the mechanism of desynchronization of the auto-oscillations can be obtained by analyzing the mappings

$$N_1(t+1) = [(1-D)N_1(t) + DN_2(t)]F(N_1), \tag{1.13}$$

$$N_2(t+1) = [DN_1(t) + (1-D)N_2(t)]F(N_2). \tag{1.14}$$

1.2. The equations of a multispecies biocenosis. Let $N_1 \in R^n$ be the number of prey, and $N_2 \in R^m$ the number of predators. We assume that under the idealized conditions of the preceding subsection the interaction of these species is described by the system of vector equations

$$\dot{N}_1 = [r_0 - A_{11}N_1 - A_{12}N_2 - a_1 N_1(t-h_1)]N_1, \tag{1.15}$$

$$\dot{N}_2 = [A_{21}N_1 - A_{22}N_2 - a_2 N_2(t-h_2)]N_2, \tag{1.16}$$

where the coordinates of the vectors r_0, a_1, and a_2 are positive, the elements of the matrices A_{11}, A_{12}, A_{21}, and A_{22} are nonnegative, the main diagonals of A_{11} and A_{22} are zero, multiplication of vectors is componentwise, $h_1 \in R^n$ and $h_2 \in R^m$ are vectors with positive coordinates, and the coordinate of h_1, h_2 corresponding to a coordinate of N_1, N_2 is subtracted from t. The system (1.15), (1.16) is structurally reminiscent of the well-known Volterra models [8]. However, there are actually three important differences.

First, (1.16) does not contain the terms called the "mortality coefficients" of the predators (it was explained in the preceding subsection that the birthrate and mortality functions are secondary).

Second, the delays enter the system (1.15), (1.16) in a completely definite way (the idea of Hutchinson is generalized).

Third, we assume that the system (1.15), (1.16) has an equilibrium state (K_1, K_2) with positive coordinates. Then it can be written in the form

$$\dot{N}_1 = r_1\left[e_n + A\left(e_n - \frac{N_1}{K_1}\right) + B\left(e_m - \frac{N_2}{K_2}\right) - \frac{N_1(t-h_1)}{K_1}\right]N_1, \tag{1.17}$$

$$\dot{N}_2 = r_2\left[C\frac{N_1}{K_1} + D\left(e_m C\frac{N_1}{K_1} - \frac{N_2}{K_2}\right) - \frac{N_2(t-h_2)}{K_2}\right]N_2, \tag{1.18}$$

where division of vectors is coordinatewise, all the coordinates of the vectors $e_n \in R^n$ and $e_m \in R^m$ are equal to 1, the coordinates of the vectors $r_1 \in R^n$ and $r_2 \in R^m$ are positive, all the elements of the $n \times n$ and $m \times m$ matrices A and D are nonnegative and their main diagonals are zero, all the elements of the $n \times m$ matrix B are nonnegative, all the elements of the $m \times n$ matrix C are nonnegative and $Ce_n = e_m$, and in multiplying through the parentheses in (1.18) D is applied to e_m first, and then the vectors are multiplied.

The system (1.17), (1.18) has the property of invariance; that is, its structure is preserved if there is a competitive displacement of some of the species. The main merit of the system (1.17), (1.18) is that all its coefficients are intrinsic characteristics of the ecosystem that have a clear biological meaning. These circumstances are of essential significance in the numerical determination of the coefficients of the system (1.17), (1.18) on the basis of an analysis of experimental data. As an example we cite [48], where an explanation is given for a fairly mysterious phenomenon—the role of the ecological trap in the stabilization of a biological community. In the solution of this problem it was not necessary to devise anything—the problem was reduced to identification of the parameters.

We underscore that the concept of Malthusian coefficients of linear growth makes sense for the prey, because they do not change upon competitive displacement of some of the species. The values of the coefficients of linear growth for the predators do change, because of a change in their food base. However, the coefficients characterizing the resistence of the environment (the coordinates of the vectors a_1, a_2 in (1.15), (1.16) and r_1/K_1, r_2/K_2 in (1.17), (1.18)) remain unchanged both for the prey and for the predators.

As an illustrative example we consider the predator-prey problem. According to (1.17), (1.18), it is modeled by the system

$$\dot{N}_1 = r_1\left[1 + a_{12}\left(1 - \frac{N_2}{K_2}\right) + b_{13}\left(1 - \frac{N_3}{K_3}\right) - \frac{N_1(t-h_1)}{K_1}\right] N_1,$$

$$\dot{N}_2 = r_2\left[1 + a_{21}\left(1 - \frac{N_1}{K_1}\right) + b_{23}\left(1 - \frac{N_3}{K_3}\right) - \frac{N_2(t-h_2)}{K_2}\right] N_2,$$

$$\dot{N}_3 = r_3\left[c_1\frac{N_1}{K_1} + c_2\frac{N_2}{K_2} - \frac{N_3(t-h_3)}{K_3}\right] N_3,$$

where $c_1, c_2 > 0$ and $c_1 + c_2 = 1$. Assume that

$$1 + a_{21} + b_{23} < (a_{21} + c_1 b_{23})\widetilde{K}_1/K_1,$$

where

$$\frac{\widetilde{K}_1}{K_1} = \frac{1 + a_{12} + b_{13}}{1 + c_1 b_{13}}.$$

Under this condition the system under consideration has a stable regime for which

$$N_1(t) > 0, \qquad N_2(t) \equiv 0, \qquad N_3(t) > 0;$$

that is, competitive displacement of the second prey is possible. If this happens, then after the transition process we drop down to the two-dimensional system

$$\dot{N}_1 = \tilde{r}_1 \left[1 + \tilde{b}_{13}\left(1 - \frac{N_3}{\widetilde{K}_3}\right) - \frac{N_1(t-h_1)}{\widetilde{K}_1}\right] N_1,$$
$$\dot{N}_3 = \tilde{r}_3 \left[\frac{N_1}{\widetilde{K}_1} - \frac{N_3(t-h_3)}{\widetilde{K}_3}\right] N_3,$$

where $\tilde{b}_{13} = c_1 b_{13}$ and $\tilde{r}_3 = c_1 r_3 \tilde{K}_1/K_1$. The last formula means that the coefficient of linear growth of the predator changes in a fairly complicated fashion; it can increase, and it can decrease. Further, by the equalities

$$\widetilde{K}_3/K_3 = c_1\widetilde{K}_1/K_1, \qquad \tilde{r}_3/\widetilde{K}_3 = r_3/K_3$$

we get in the limit as $c_1 \to 0$ that

$$\dot{N}_3 = -r_3\frac{N_3(t-h_3)}{K_3}N_3;$$

that is, the predator dies out together with the second prey, since the first prey ceases to be a part of the predator's food ration.

If the area in which events develop has a pronounced spatial heterogeneity, then we must take migration factors into account. Here the general structure of equations (1.17), (1.18) is preserved [42], except that on their right-hand sides we must add the respective terms

$$\frac{1}{K_s}\frac{\partial}{\partial x_1}\left(K_s^2 D_s \frac{\partial}{\partial x_1}\frac{N_s}{K_s}\right) + \frac{1}{K_s}\frac{\partial}{\partial x_2}\left(K_s^2 D_s \frac{\partial}{\partial x_2}\frac{N_s}{K_s}\right),$$

where $s = 1, 2$, and set the boundary conditions of impenetrability. But if a temporal heterogeneity is also added, then the ecological equations become indeterminate to a significant degree, because it is necessary to work with equations (1.15), (1.16), for which it is difficult to identify the parameters.

It is interesting that in all the cases investigated the role of spatial heterogeneity is the same—it is an additional stabilizing factor, since it effectively promotes suppression of the oscillations.

This situation leads to the following conclusions in connection with the human population. As a biological species in the process of evolution, the human race has lost stability, a circumstance connected mainly with changes in the age structure, while complexification in the social structure has promoted a certain stabilization. It turns out that if the resistance of the external environment is multidimensional and depends in a sufficiently smooth way on the space variables, then the oscillations are completely suppressed when the remaining parameters are biologically reasonable. Thus, on a certain level there is stabilization of the size of the human population, whose growth is conditioned by technological progress, only if the infrastructure in each government is complexified in a suitable way, the legal and social protection of the individual person is strengthened, work conditions and rest conditions are improved, etc. We remark that we have excluded from consideration the case of degeneration of a nation, which has been observed, for example, in Russia during the past two decades of stagnation, when there has even been a decrease in the population.

Of course, many subtleties have been left out in such a brief exposition, in particular, the connection between the model (1.17), (1.18) and the ideas of the most prominent biologists about the reasons for oscillations in the numbers of species in nature (for this we refer to the monograph [54]).

1.3. The stabilizing role of the ecological trap. This problem was mentioned above. Its source was the article [3], which describes the following ecological situation that has taken shape on the island of Newfoundland. The number of species dwelling on this island is not large. The biggest of the abundant predators is the lynx. Its main prey are the white hare, the caribou, and the arctic hare. A typical feature of the functioning of this simple ecosystem is as follows. It is clear that the arctic hare is poorly adapted to defense against the predator, and most of its encounters with the lynx therefore end with its death. The low density of the arctic hare thus increases only in times of a rise in the density of the white hare and the caribou, when the lynx feeds mainly on these two species. In this connection it is stated in [3] that the arctic hare finds itself in an ecological trap. In [3] it is also noted that the ecological trap plays an important stabilizing role. This involves the lynx making a felicitous selection of prey and having the ability to switch from one prey to another.

Thus, it can be assumed that the deeper essence of the ecological trap is not revealed in [3]. Our analysis shows that it is very difficult to do this without resorting to mathematical modeling.

Based on the last two subsections, we conclude that the dynamics of the ecosystem described above is modeled by the following differential equations

(1.19)
$$\dot{N}_1 = r_1[1 + a_{14}(1 - N_4) + b_{12}(1 - N_2) = \alpha_{11}N_1(t - h_1) - \alpha_{12}N_1(t - h_1 - 1)]N_1,$$
(1.20)
$$\dot{N}_2 = r_2[1 + a_{24}(1 - N_4) + b_{21}(1 - N_1) - \alpha_{21}N_2(t - h_2) - \alpha_{22}N_2(t - h_2 - 1)]N_2,$$
(1.21)
$$\begin{aligned}\dot{N}_3 = r_3[1 &+ a_{34}(1 - N_4) - \alpha_{31}N_3(t - h_3) \\ &- \alpha_{32}N_3(t - h_3 - 1) - \alpha_{33}N_3(t - h_3 - 2) - \alpha_{34}N_3(t - h_3 - 3)]N_3,\end{aligned}$$
(1.22)
$$\begin{aligned}\dot{N}_4 = r_4[c_1N_1 &+ c_2N_2 + c_3N_3 - \alpha_{41}N_4(t - h_4) \\ &- \alpha_{42}N_4(t - h_4 - 1) - \alpha_{43}N_4(t - h_4 - 2) - \alpha_{44}N_4(t - h_4 - 3)]N_4.\end{aligned}$$

Here N_1, N_2, N_3, and N_4 are the densities of the white hare, the arctic hare, the caribou, and the lynx, respectively. Further, in (1.19)–(1.22)

$$r_1 = \frac{\lambda_1}{1 + a_{14} + b_{12}}, \qquad r_2 = \frac{\lambda_2}{1 + a_{24} + b_{21}}, \qquad r_3 = \frac{\lambda_3}{1 + a_{34}}, \tag{1.23}$$

where λ_1, λ_2, and λ_3 are the Malthusian coefficients of linear growth.

We describe a way of choosing the numerical values of the parameters in the system (1.19)–(1.22). The intrinsic characteristics of these species are determined in a very simple way: the Malthusian coefficients of linear growth, the ages of sexual maturity of females, the number of age groups, and the coefficients characterizing them. See the table.

Species	Malthusian coefficient of linear growth	Age of sexually mature females	Number of age groups	Characteristic of age structure
White hare	1.44	1.86	2	0.6; 0.4
Artic hare	0.92	1.86	2	0.9; 0.1
Caribou	0.41	1.37	4	0.28; 0.26; 0.24; 0.22
Lynx	0.35	1.81	4	0.5; 0.27; 0.15; 0.08

We give brief explanations. The Malthusian coefficients of linear growth of the white hare and the caribou are obtained with the help of (1.4) and by analyzing the experimental material in [17] and [66]. The corresponding coefficient for the arctic hare was actually given in [3]. The coefficient of linear growth of the lynx was determined with the biological material in [57] and [66] taken into account. The age of sexual maturity of females is computed without difficulty. The number of age groups and their numerical characteristics were chosen for the white hare according to [66]. The corresponding parameters for the caribou were obtained by analyzing the age composition of the Bulun herd of wild reindeer [17], which is regarded as standard. However, these parameters for the lynx were chosen in a rather arbitrary way, because there is no corresponding data in the literature [57]. The same remark applies to the arctic hare.

We now turn to the coefficients characterizing interaction between the species. Data on ordinary losses from predation [66] allow us to assume that

$$a_{14} = 0.1\,. \tag{1.24}$$

It is indicated in [3] that from May to October more than 70% of the young caribou die as a result of lynx attack. In view of (1.23) and the table, this implies that the coefficient of the pressure of the lynx on the caribou is

$$a_{34} = 1.9\,. \tag{1.25}$$

We pass to the parameters c_1, c_2, and c_3—the fractions of each prey in the ration of the predator. As mentioned above, a young caribou is easy prey for the lynx. It is natural to assume that during the corresponding time the caribou calves constitute the main food supply for the lynx. Thus, the relative duration of the interval from May to October is taken as the coefficient c_3; that is,

$$c_3 = 0.35\,. \tag{1.26}$$

We remark in addition that in those national preserves where the lynx flourishes the fraction of caribou calves in its ration is roughly the same [57]. The fraction of the arctic hare in the ration of the lynx was estimated from approximate calculations of the size of the lynx population, the amount of food needed by a single lynx, and the consideration that the predator annihilates almost all the offspring of the arctic hare, the average number of which is fairly stable on the island (about a thousand). From estimates of this kind it appears that the arctic hare constitutes several percent of the total catch of the lynx. Below it is assumed that

$$c_2 = 0.05\,. \tag{1.27}$$

From this it automatically follows that

$$c_1 = 0.6\,. \tag{1.28}$$

These number do not change in what follows.

It remains to choose the coefficients of the competitive struggle between the two species of hare and the coefficient of the pressure of the lynx on the arctic hare. In [3] it is mentioned that all else being equal, except in the absence of the lynx, the arctic hare adapts and propagates itself well. Therefore, the competitive struggle between these two species of hares is not a hard one. In this case the competition coefficients are not large. Below we take

$$b_{21} = 0.2\,; \tag{1.29}$$

that is, we assume that the coefficient of the pressure of the white hare on the arctic hare is rather noticeable. Its precise value is inessential. Considerably more important is the coefficient b_{12} of competition from the side of the arctic hare, which we will vary below in a certain range. As mentioned at the beginning of this subsection, the coefficient a_{24} of the pressure of the lynx on the arctic hare is for biological reasons fairly large. We first assume that

$$a_{24} = 5\,. \tag{1.30}$$

Figures 1.3–1.6 show the dynamics of the change in the stationary regimes of the system (1.19)–(1.22) with the respective values $b_{12} = 0.1$, 0.16, 0.3, and 0.4 for the values in the table and for the data (1.24)–(1.30).

It is clear that the oscillations stabilize as b_{12} increases. What is more, it turns out that for $b_{12} \geq 0.5$ the equilibrium state of the system (1.19)–(1.22) with coordinates equal to 1 becomes stable. The biological meaning of this is transparent: the oscillations of the densities of all the species depend in a very essential way on the degree of intensity of the oscillations in the density of the white hare, which decrease as the coefficient b_{12} of competition from the side of the arctic hare increases. Here the coefficient a_{24} of the pressure of the lynx on the arctic hare must necessarily be sufficiently large. To see this, it suffices to compare Figures 1.5 and 1.7, where in the latter case it is assumed that $a_{24} = 3$ instead of (1.30). We mention also that for $a_{24} > 5$ the graphs shown in Figures 1.3–1.6 change little.

As we mentioned above, some of the parameters in the system (1.19)–(1.22) were chosen with a large error. It is essential that on a qualitative level this does not affect the dynamics of the stationary regimes of the system (1.19)–(1.22).

Accordingly, it became clear from an analysis of the mathematical model (1.19)–(1.22) that the central role in the stabilization of the ecosystem under consideration is played by two circumstances: the considerable pressure of the lynx on the arctic hare, and the successful competition of the latter with the white hare. To make everything finally clear, we emphasize that the cause of instability in this ecosystem is the excessively intense oscillations in the density of the white hare. But the simultaneous

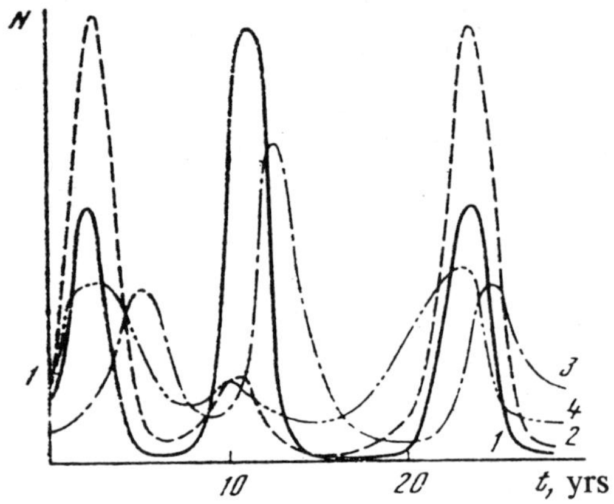

FIGURE 1.3

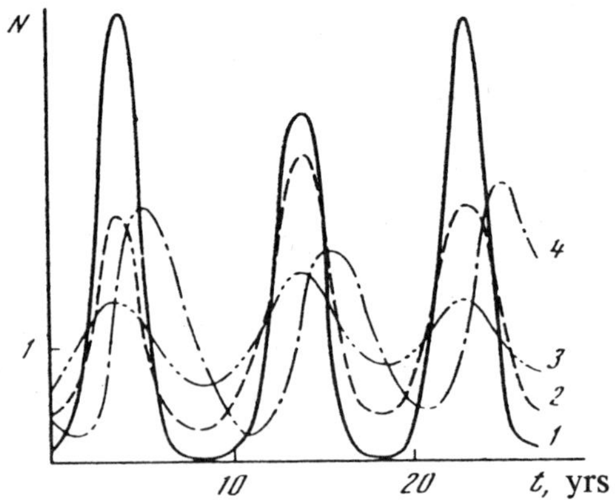

FIGURE 1.4

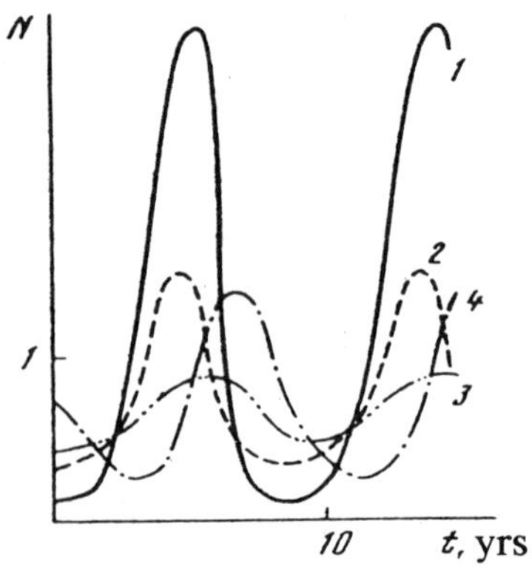

FIGURE 1.5

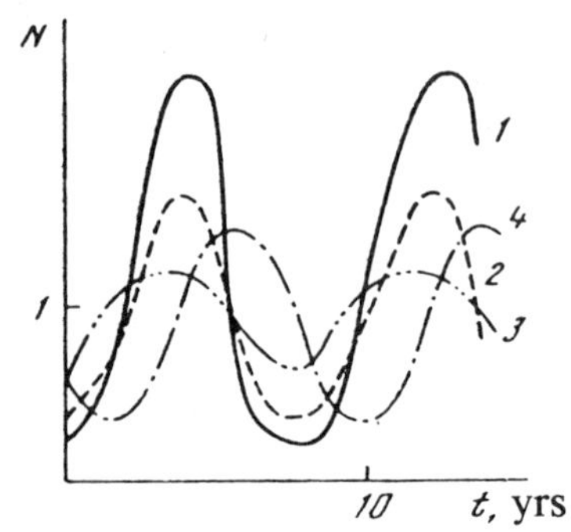

FIGURE 1.6

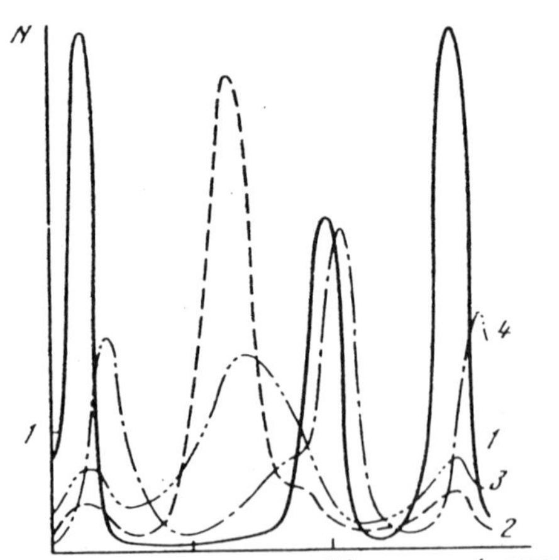

FIGURE 1.7

action of the two causes indicated above leads to the lynx being able to effectively wipe them out. The essence of the ecological trap is thus the following: the lynx exerts marked pressure on the white hare in an indirect way—by strongly oppressing the population of the arctic hare.

The example considered in this subsection is interesting also in that it clarifies the nature of the so-called ten-year cycle of oscillations in the density of mammals in the taiga zones of Russia and Canada (see [66] and [79]). A three-year cycle is typical for their tundra regions, as conditioned by the magnitude of the period of oscillations in the density of the lemming, which is a basic species in these areas.

1.4. A mathematical model of diabetes. We begin with a brief exposition of the mechanism involved in the process of regulation of the glucose level in the blood. A person usually throws off excess sugar. Its concentration is lowered by active insulin, which is a derivative of the insulin manufactured by the pancreas. The working intensity of the latter increases as the concentration of glucose in the blood increases, and decreases as the concentration of active insulin increases. The concentration of active insulin depends also on the amount of glucose in the blood.

This brief description suffices to lead to the following conclusion. The insulin is rightfully regarded as the first-order predator N_1, the active insulin as the second-order predator N_2, and the glucose as the prey N_3. A distinction of the situation here is that the prey N_3 can stimulate or suppress the growth of the predator N_2.

If we use our modeling method, we are therefore led to the system

$$\dot{N}_1 = r_1[N_3/K_3 + a_{12}(N_3/K_3 - N_2/K_2) - N_1(t-h)/K_1]N_1, \tag{1.31}$$

$$\dot{N}_2 = r_2[N_1/K_1 + a_{23}(N_1/K_1 - N_3/K_3) - N_2/K_2]N_2, \tag{1.32}$$

$$\dot{N}_3 = r_3[1 + a_{32}(1 - N_2/K_2) - N_3/K_3]N_3, \tag{1.33}$$

where only the parameter a_{23} has arbitrary sign: for $a_{23} < 0$ the glucose promotes the production of active insulin, while for $a_{23} > 0$ it suppresses it. By analogy with the case of the ecological equations it is reasonable to assume that the quantities

$$r_3(1 + a_{32}), \quad r_1/I_1, \quad r_2/K_2, \quad r_3/K_3 \tag{1.34}$$

are invariant; that is, they are preserved under any change in the regulation parameters a_{12}, a_{23}, and a_{32}. In (1.31) a delay of $h = 2.5$ to 3.5 hours characterizes the average time needed for the pancreas to manufacture insulin.

It is simple to deduce from the most superficial analysis of the characteristic quasipolynomial of the system linearized on the equilibrium state (K_1, K_2, K_3) that there is oscillatory instability in a broad range of variation of the parameters. In this case a periodic solution with period roughly equal to 12 hours is a stable regime of the system (1.31)–(1.33). We list some nontrivial properties of this cycle.

(a) Suppose that K_1, K_3, and a_{23} are relatively large, and K_2, a_{12}, and a_{32} are small. Under these conditions N_3 oscillates rather intensely. This case models diabetes.

(b) Suppose that all the parameters have moderate values and $a_{23} < 0$. Under these conditions the oscillations of N_3 are also moderate. This case corresponds to the normal situation.

(c) Suppose that a_{23} and a_{32} are large, while K_3 and a_{12} are small. The oscillations of N_3 are sharpest under these conditions. This case models a fairly rare disease when the average glucose level in the blood is excessively low.

These properties of the system (1.31)–(1.33) can be understood without resorting to numerical integration of it if we use the postulated property of constancy of the quantities in (1.34). In fact, we allowed ourselves to present the system (1.31)–(1.33) in order to stress the role of the invariants; this system was obtained by the second author back in 1983. The material presented here was reported by him in a seminar on the theory and applications of differential equations at Yaroslavl University.

1.5. Explanation of the "plankton paradox". This question is connected with competition of species with strongly intersecting requirements on the surrounding environment. For simplicity we take the case of three species. Then

$$\dot N_1 = r_1[1 + a_{12}(1 - N_2) + a_{13}(1 - N_3) - N_1(t - h_1)]N_1, \tag{1.35}$$

$$\dot N_2 = r_2[1 + a_{21}(1 - N_1) + a_{23}(1 - N_3) - N_2(t - h_2)]N_2, \tag{1.36}$$

$$\dot N_3 = r_3[1 + a_{31}(1 - N_1) + a_{32}(1 - N_2) - N_3(t - h_3)]N_3. \tag{1.37}$$

Let

$$\alpha_1 = a_{31}\frac{\lambda_1}{r_1} - \frac{\lambda_3}{r_3}, \qquad \alpha_2 = \frac{\lambda_2}{r_2} - a_{21}\frac{\lambda_1}{r_1}, \tag{1.38}$$

$$\alpha_3 = a_{12}\frac{\lambda_2}{r_2} - \frac{\lambda_1}{r_1}, \qquad \alpha_4 = \frac{\lambda_3}{r_3} - a_{32}\frac{\lambda_2}{r_2}, \tag{1.39}$$

$$\alpha_5 = a_{23}\frac{\lambda_3}{r_3} - \frac{\lambda_2}{r_2}, \qquad \alpha_6 = \frac{\lambda_1}{r_1} - a_{13}\frac{\lambda_3}{r_3}, \tag{1.40}$$

where λ_1, λ_2, and λ_3 are the Malthusian coefficients of linear growth.

THEOREM 1.1. *Suppose that the nonzero numbers in* (1.38)–(1.40) *have the same sign. Then the system* (1.35)–(1.37) *has as an invariant set a homoclinic triangle lying entirely on the boundary of the cone of vector-valued functions with nonnegative components.*

In Figure 1.8 we show this homoclinic triangle for the case when the system (1.35)–(1.37) does not have delays.

We introduce the notation $\alpha = \alpha_1\alpha_3\alpha_5/\alpha_2\alpha_4\alpha_6 - 1$.

THEOREM 1.2. *Assume the conditions of Theorem* 1.1. *Then for* $\alpha > 0$ (< 0) *the homoclinic triangle is locally exponentially stable* (*unstable*).

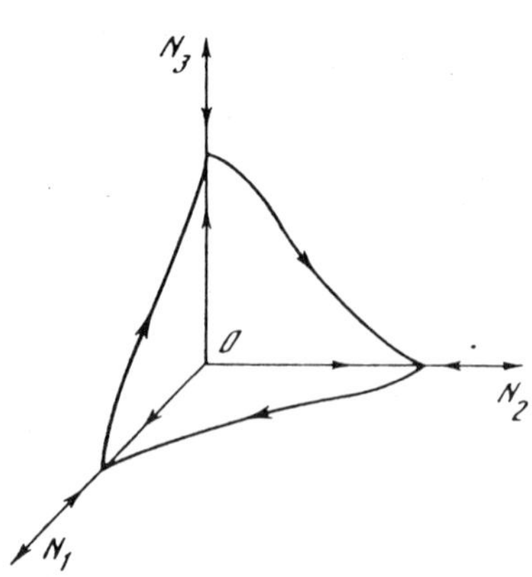

FIGURE 1.8

As an illustrative example we consider the system

$$\dot{N}_1 = \frac{\lambda}{1+a}[1 + a(1 - N_3) - N_1(t-1)]N_1, \tag{1.41}$$

$$\dot{N}_2 = \frac{\lambda}{1+a}[1 + a(1 - N_1) - N_2(t-1)]N_2, \tag{1.42}$$

$$\dot{N}_3 = \frac{\lambda}{1+a}[1 + a(1 - N_2) - N_3(t-1)]N_3. \tag{1.43}$$

In the system (1.41)–(1.43) a homoclinic triangle arises for $a > 1$, becoming stable for $a > 2$. We explain its biological meaning: through a hard competitive struggle the first species (say) initially almost displaces the remaining two species, then a partial victory passes to the third species over a sufficiently long period of time, and then to the second, and so on. It is essential that in the course of time the local minima of the density of each of the species are deepened. This means that only some one species survives.

To the right-hand sides of the system (1.35)–(1.37) we add the diffusion terms and substitute the impenetrability conditions on the boundary of some domain.

THEOREM 1.3. *Under the conditions of Theorem* 1.1 *suppose that* $\alpha > 0$. *Then for a suitable decrease in the diffusion coefficients the homoclinic triangle is unstable in the framework of the distributed model.*

The proofs of Theorems 1.1 and 1.2 are carried out according to the schemes in [37], with appropriate changes connected with the presence of a delay in the system (1.35)–(1.37). We do not yet have a rigorous proof of Theorem 1.3. However, from a biological point of view the corresponding assertion is obvious; therefore we formulated it as a theorem. We mention further that Theorem 1.3 ceases to be true in general if at least one of the delays in (1.35)–(1.37) vanishes.

A numerical analysis shows that the simplest scenario is as follows. For small competition coefficients the components of an attractor of the system (1.35)–(1.37) are bounded away from zero. Then as the hardness of the competitive struggle increases, the attractor attaches to the homoclinic triangle, which becomes a global attractor. In more complex scenarios an attractor of ordinary type can coexist with a stable homoclinic triangle. In all variants, however, taking the migration factor into account leads to the situation in which, instead of a homoclinic triangle, a stationary regime with components bounded away from zero becomes an attractor.

We apply the above facts to explain the well-known "plankton paradox". It is essentially the following. More than fifty years ago Gause formulated the proposition that species with strongly intersecting requirements on the surrounding environment cannot coexist. But simultaneously existing species of plankton are just such species. As we have shown, the paradox is resolved if together with delays we take into account a small migration factor, as is natural for the given biological communities.

2. Properties of a relaxation cycle of the Hutchinson equation

2.1. Statement of the problem and a theorem on the existence of a relaxation cycle. Normalizing the time $t \to ht$ in equation (1.3) and setting $\lambda = rh$, we get

$$\dot{N} = \lambda[1 - N(t-1)]N. \tag{2.1}$$

We ask about the asymptotic behavior and stability of a relaxation cycle of equation (2.1) under the assumption that λ is a large parameter. As its phase space we take $C(-2, -1)$. Denote by S_0 the closed convex subset of $C(-2, -1)$ whose elements are the functions

$$0 \le N(t, \lambda) \le \exp[m\lambda(t+1)], \qquad N(-1, \lambda) = 1, \tag{2.2}$$

where the positive constant m is less than 1.

THEOREM 2.1. *There is a sufficiently large positive number λ_0 such that for each $\lambda \ge \lambda_0$ the first-return operator Π of equation* (2.1) *is defined on the set S_0 and transforms it into itself.*

Integrating equation (2.1) by steps with the initial conditions (2.2), we see that its solutions on the interval $[-1, 0]$ increase to exponentially large values. Upon passing through zero each solution falls in an asymptotically short time to superexponentially small values, and then its growth goes an asymptotically long time roughly according to the Malthusian law. Therefore, there is a first value $T-1$ of the time such that $N(T-1, \lambda) = 1$ and $\dot{N}(T-1, \lambda) > 0$. Let

$$\Pi N(t, \lambda) = N(T+t, \lambda), \qquad -2 \le t \le -1.$$

It follows from the construction that the first-return operator is defined on S_0 and transforms it into itself.

2.2. Asymptotics of a relaxation cycle. It follows from the proof of Theorem 2.1 that a periodic solution grows for a long time before falling into S_0. It follows from this and the structure of (2.1) that

$$N(t, \lambda) = [1 + O(\exp(-2\lambda))]c \exp \lambda(t+1), \tag{2.3}$$

where $-3 \le t \le -2$, and the constant c has order 1. We deduce from (2.3) and (2.1) that

$$N(t, \lambda) = [1 + \exp(-\lambda) + O(\exp(-2\lambda))] \exp\{\lambda(t+1) - \exp \lambda t\}, \qquad -2 \le t \le -1. \tag{2.4}$$

In turn, we get from (2.4) and (2.1) that

$$\begin{aligned} N(t, \lambda) = [1 &+ \exp(-\lambda) - \exp \lambda(t-1) \\ &+ \tfrac{1}{2} \exp \lambda(2t-1) + O(\exp(-2\lambda))] \exp\{\lambda(t+1) - \exp \lambda t\}, \\ &- 1 \le t \le 0. \end{aligned} \tag{2.5}$$

Denote the largest value of $N(t, \lambda)$ by $\overline{N}$. By (2.2) and (2.1), $\overline{N} = N(0, \lambda)$. From this and (2.5) we conclude that

$$\overline{N} = \exp(\lambda - 1) + \tfrac{1}{2}e + O(\exp(-\lambda)). \tag{2.6}$$

We proceed to the interval $[0, 1]$. Using (2.5) in (2.1), we first refine it so that we then get the equality

$$\begin{aligned} N(t, \lambda) = \exp\{\lambda(t+1) &- \exp \lambda + \exp[\lambda - \exp \lambda(t-1)] \\ &+ \tfrac{1}{2} \exp[2\lambda(t-1) - \exp \lambda(t-1)] \\ &+ \exp[-\exp \lambda(t-1)] - 1 + \exp(-\lambda) + O(\exp(-2\lambda))\}. \end{aligned} \tag{2.7}$$

An important characteristic is the value of the time $t_0 > 0$ when, decreasing sharply for the first time, the periodic solution reaches the value 1. We conclude from (2.7) that

$$t_0 = \ln\lambda/(\lambda - 1) + O(\ln^2\lambda/\lambda^3). \tag{2.8}$$

We can now proceed to the construction of the asymptotics of the cycle period. On the interval $[1, T-4]$ we replace $N(t, \lambda)$ first by zero, and then by $3\exp(-3\lambda)$. After this we recall that, according to (2.1), a periodic solution has mean value 1. Therefore, the device described above with change of $N(t, \lambda)$ enables us to estimate T from above and from below. This and (2.3)–(2.5) and (2.7) give us

$$T = (1 + \exp\lambda)/\lambda + O(\exp(-\lambda)/\lambda). \tag{2.9}$$

It follows from (2.1) that the smallest value $\underline{N}$ of the periodic solution $N(t, \lambda)$ is attained at the time $1 + t_0$. It is thus necessary by using (2.7) to integrate equation (2.1) a little further, which leads to the equality

$$\underline{N} = \exp\left[-\exp\lambda + 2\lambda - 1 + \frac{1 + (1+\lambda)\ln\lambda}{\lambda} + O\left(\frac{\ln^2\lambda}{\lambda^2}\right)\right]. \tag{2.10}$$

Computing according to (2.6), (2.9), and (2.10) and comparing with Figure 1.1, we conclude that the asymptotic formulas work already for $\lambda \geq 2.5$.

In applications to population dynamics of mammals we encounter cases in which the parameter λ is relatively small; that is, our asymptotic theory is not suitable. However, it is applicable to populations of fish with sufficiently large Malthusian coefficient of linear growth (which can be measured with difficulty and varies from area to area). In particular, (2.10) makes it clear why single-species ichthyocenoses are practically absent in nature. An exception is formed by species that in the evolution process have developed special mechanisms for suppressing oscillations, for example, partial spawning in the case of the carp family or cannibalism in the case of the perch family [53].

We dwell very briefly on the properties of a relaxation cycle of the equation

$$\dot{N} = \lambda[1 - \alpha N(t-1) - (1-\alpha)N(t-1-\gamma)]N, \tag{2.11}$$

where $0 < \alpha < 1$ and $\gamma > 0$ are fixed parameters, and λ is a large parameter. We remark that (2.11) is the simplest variant of equation (1.5). The assertion of Theorem 2.1 is preserved for (2.11), but the asymptotic formulas for a relaxation cycle change. Without dwelling on this question in detail, we note only that $\overline{N}$ and T become roughly $1/\alpha$ times larger, and $\underline{N}$ roughly just as many times smaller. In the language of biology these facts reflect the obvious circumstance that long-lived individuals are not needed by a strongly fertile species. Incidentally, we recall that this conclusion was actually drawn in §1.1—in the case of a strongly fertile species oscillations were noticeable only in the youngest age group; that is, there were practically no other age groups.

2.3. Stability and uniqueness of a relaxation cycle. In the variational equation

$$\dot{v} = -\lambda N(t, \lambda)v(t-1) + \frac{\dot{N}(t, \lambda)}{N(t, \lambda)}v \tag{2.12}$$

of a relaxation cycle $N(t, \lambda)$ we make the normalizing substitution

$$v_1 = v/N(t, \lambda) \tag{2.13}$$

and pass to the associated equation

$$\dot{w} = \lambda N(t, \lambda) w(t+1). \tag{2.14}$$

It is convenient to take $C(2, 3)$ as the phase space of the latter. The following character of behavior of a solution $w(t, \lambda)$ of equation (2.14) with arbitrary initial condition in $C(2, 3)$ is a consequence of an analysis carried out in the next section. First, everything is clear: for $1 \le t \le 2$ the function $w(t, \lambda)$ is close with superexponential accuracy to a constant w_0—the value of $w(t, \lambda)$ at $t = 2$. It is also clear that for $0 \le t \le 1$ an essential change of $w(t, \lambda)$ begins only in an asymptotically small neighborhood of zero. It continues to change strongly for $-1 \le t \le 0$. But already for $-2 \le t \le -1$ there is a stabilization of its values—with high accuracy it becomes close to the same constant w_0.

To finally clarify the situation we consider the Hale inner product [35]

$$w(t, \lambda)\dot{N}(t, \lambda)/\lambda N(t, \lambda) - \int_{t-1}^{t} N(s, \lambda) w(s+1, \lambda)\, ds \tag{2.15}$$

of the functions $\dot{N}(t, \lambda)/\lambda N(t, \lambda)$ and $w(t, \lambda)$. Recall that (2.15) does not really depend on t. It is convenient to take, say, $t = -2$. Then it follows from the asymptotic properties of the cycle $N(t, \lambda)$ and of the solutions $w(t, \lambda)$ of (2.14) that (2.15) tends to w_0 as $\lambda \to \infty$.

We summarize the obvious conclusions. As $\lambda \to \infty$, 1 is a simple multiplier of the monodromy operator of equation (2.12). Further, the rest of its multipliers tend to zero. Therefore, we have the following assertion.

THEOREM 2.2. *The first-return operator* Π *constructed in the proof of Theorem* 2.1 *has a unique fixed point in* S_0, *and the corresponding relaxation cycle of equation* (2.1) *is orbitally exponentially stable.*

Formally, equation (2.1) does not have a limit object as $\lambda \to \infty$. However, Theorems 2.1 and 2.2 in combination allow us to regard its relaxation cycle as such an object. Therefore, we can draw a parallel between (2.1) and the van der Pol equation with a large parameter in the classical theory of relaxation oscillations [62].

2.4. Another way of investigating the variational equation. We present a different method for analyzing the behavior of solutions of (2.14) as $t \to -\infty$, a method that will be useful in what follows. Making the substitution

$$w = w_1 \exp \lambda t, \tag{2.16}$$

we pass from (2.14) to the equation

$$\dot{w}_1 = -\lambda w_1 + \mu a(t, \lambda) w_1(t+1), \tag{2.17}$$

in which

$$a(t, \lambda) = \lambda N(t, \lambda) \exp \lambda. \tag{2.18}$$

Moreover, the parameter $0 \le \mu \le 1$ now appears in (2.17). The meaning of its introduction will soon become clear.

It will be assumed that in (2.17) the function $a(t, \lambda)$ is given by (2.18) only for $1 \leq t \leq T-3$, where T is the period of $N(t, \lambda)$. For $T-3 \leq t \leq T-3+\delta$, where the parameter $\delta > 0$ is arbitrarily small, we change the function $a(t, \lambda)$ by monotone decrease so that

$$a(T-3+\delta, \lambda) = a(1, \lambda),$$

and then we extend it to the whole number line by periodicity. It then follows from the smallness of $a(t, \lambda)$ and results in [32] and [47] that there is an exponential dichotomy for the solutions of the modified equation (2.17). It is essential that the norms of the projections implementing it are bounded uniformly with respect to λ and μ. Further, it is completely obvious that for $\mu = 0$ the subspace of solutions of (2.17) that increase exponentially as $t \to -\infty$ is one-dimensional, and hence it will also be one-dimensional for $\mu = 1$. From this and (2.16) it follows that

$$\|U(1, T-2)w_1(t)\| \leq M_1 \exp(-\gamma_1 \exp\lambda)\|w_1(t)\|. \tag{2.19}$$

Here M_1 and γ_1 are universal positive constants, $U(t, \tau)$ is the Cauchy operator of equation (2.14), and $w_1(t)$ belongs to a suitable subspace of $C(T-3, T-2)$ of codimension 1. Integrating (2.14) by steps, we see that

$$\|U(-2, 1)\| \leq M_2 \exp\gamma_2\lambda, \tag{2.20}$$

where M_2 and γ_2 are positive constants. We conclude from a comparison of (2.19) and (2.20) that the solutions of (2.14) with initial conditions in the above subspaces of $C(T-3, T-2)$ become superexponentially small in the time T. Further, it is obvious that the solutions with initial conditions in the complementary one-dimensional subspace are periodic solutions of (2.14).

2.5. The stabilizing role of cannibalism. Biologically, this is a situation when inviduals of the oldest age group eat the young. In the simplest case this therefore involves the equation

$$\dot{N} = \frac{\lambda}{1+a}[1 + a(1-N) - N(t-1)]N, \tag{2.21}$$

where the positive parameter a characterizes the intensity of cannibalism (the size of the population is decreased by a factor of $1+a$ as a result of cannibalism). Linearizing equation (2.21) on its single equilibrium state, we get

$$\dot{v} = -\frac{a\lambda}{1+a}v - \frac{\lambda}{1+a}v(t-1). \tag{2.22}$$

LEMMA 2.1. *If $a > 1$, then for any λ the solutions of* (2.2) *damp out exponentially. But for $a < 1$, the analogous statement is true only when*

$$\lambda < \omega_0\sqrt{(1+a)/(1-a)}, \qquad \cos\omega_0 = -a, \ \pi/2 < \omega_0 < \pi.$$

We omit the proof, which is fairly trivial.

Thus, for $a < 1$ the single equilibrium state of equation (2.21) loses stability in an oscillatory way as λ increases.

THEOREM 2.3. *Suppose that*

$$a < 1 \quad \textit{and} \quad \lambda > \omega_0\sqrt{(1+a)/(1-a)}. \tag{2.23}$$

Then equation (2.21) *has a cycle* $N(t, \lambda)$ *such that*

$$0 < N(t, \lambda) < (1+a)/a. \tag{2.24}$$

For the proof it is convenient to take $R \times L(-2, -1)$ as the phase space of equation (2.16). In it we consider the closed convex set

$$S_0 = \{N(t) : 0 \le N(t) \le 1, N(-1) = 1\}.$$

Since all the roots of the characteristic quasipolynomial of (2.22) are complex, the first-return operator Π along trajectories of equation (2.21) is defined on S_0. Moreover, $\Pi S_0 \subset S_0$, as follows from the nonoscillation of its solutions with respect to 1. Next, we consider a root $\tau + i\sigma$ of the quasipolynomial with largest real part. From the equalities

$$\tan\sigma = -\frac{\sigma}{\tau + a\lambda/(1+a)}, \qquad \left(\tau + \frac{a\lambda}{1+a}\right)^2 + \sigma^2 = \frac{\lambda^2}{(1+a)^2}\exp(-2\tau)$$

it follows that $\pi/2 < \sigma < \pi$. From this we conclude that for $N(t) \in S_0$ and $\|1 - N(t)\| = \nu$, where the constant $\nu > 0$ is sufficiently small, the integral

$$\int_{-2}^{-1} (1 - N(t)) \sin\sigma(t+1) \exp\tau(t+1)\, dt \tag{2.25}$$

is negative uniformly with respect to the $N(t)$ under consideration. But then the fixed point $N(t) \equiv 1$ of the operator Π on S_0 has the repulsion property. The latter means the following. Suppose that $N(t) \in S_0$ and $0 < \|1 - N(t)\| < \delta$, where the constant δ is sufficiently small. Then there is an integer $m = m[N(t)]$ such that

$$\|1 - \Pi^m N(t)\| > \delta.$$

Therefore [73], the operator Π has a fixed point not identically equal to 1 in S_0. In passing we have cleared up the question of the choice of phase space: the necessary negativity of the integral in (2.25) is guaranteed.

We point out that under initial conditions in $C(-2, -1)$ satisfying the restrictions (2.2), equation (2.21) begins to work like a relay as λ increases: the main time is spent by the solutions either in a neighborhood of zero or in a neighborhood of $(1+a)/a$. In particular, the first-return operator Π is defined on the set S_0 in §2.1 and transforms this set into itself. Therefore, for large λ and $a < 1$ equation (2.21) has a relaxation cycle $N(t, \lambda)$, which obviously satisfies the inequalities (2.24), and its period is close to $(1+a)/a$. Questions involving its asymptotics and stability will be studied in Chapter III.

3. Relaxation oscillations in a bilocal model of the Hutchinson equation with diffusion

3.1. Statement of the problem. For large values of λ the equation (2.1) discussed above describes the dynamics of variation in the density of a strongly fertile population living in a homogeneous area when the food base is regularly replenished up to some fixed level, and the migration factor is so large that spatial perturbances are suppressed. Under precisely these biological assumptions we consider two areas joined by a relatively narrow passage. As a result we arrive at the system

$$\dot N_1 = D(N_2 - N_1) + \lambda[1 - N_1(t-1)]N_1, \tag{3.1}$$

$$\dot N_2 = D(N_1 - N_2) + \lambda[-N_2(t-1)]N_2, \tag{3.2}$$

where the diffusion coefficient D characterizes the depth of the connection between the areas. It is biologically obvious that if D is not too small, then the oscillations in the system (3.1), (3.2) are synchronized; that is, the unique attractor of the system is the homogeneous cycle

$$(N(t, \lambda), N(t, \lambda)), \tag{3.3}$$

which was constant in the preceding section, and the oscillations begin to desynchronize under a certain decrease in D. Our goal is to carry out a detailed analysis of the dynamics of solutions of the systems (3.1), (3.2) with positive coordinates by combining intuitive biological ideas and some special mathematical constructions.

3.2. The neutral curve. We linearize the system (3.1), (3.2) on the homogeneous cycle (3.3) and consider the projection

$$\dot{v} = -zv - \lambda N(t, \lambda)v(t-1) + \frac{\dot{N}(t, \lambda)}{N(t, \lambda)}v, \tag{3.4}$$

where $z = 2D$, on the unique spatially heterogeneous mode. In (3.4) we make the normalizing substitution (2.13) and pass to the associated equation

$$\dot{w} = zw + \lambda N(t, \lambda)w(t+1). \tag{3.5}$$

As in the preceding section, we take $C(2, 3)$ as its phase space. It is convenient to make the substitution

$$w \to w \exp zt \tag{3.6}$$

in (3.5), which leads to the equation

$$\dot{w} = \mu N(t, \lambda)w(t+1), \tag{3.7}$$

where $\mu = \lambda \exp z$. It follows from the asymptotic formulas for $N(t, \lambda)$ obtained earlier that with exponential accuracy a solution $w(t)$ of equation (3.7) with arbitrary initial condition in $C(2, 3)$ is close to the value w_0 of the initial condition at $t = 2$ up to the moment of time $\delta = \ln 3\lambda/\lambda$. Using the step method, we give a formula for determining $w_{\text{out}} = w(-3+\delta)$:

$$\begin{aligned} w_{\text{out}} = w_0 \Bigg[1 - \mu \int_{-3+\delta}^{\delta} N\, ds + \mu^2 \int_{-3+\delta}^{-1+\delta} N \int_{s+1}^{\delta} N\, d\tau\, ds \\ - \mu^3 \int_{-3+\delta}^{-2+\delta} N \int_{s+1}^{-1+\delta} N \int_{\sigma+1}^{\delta} N\, d\tau\, d\sigma\, ds\Bigg]. \end{aligned} \tag{3.8}$$

Computing the integrals in (3.8) asymptotically, we get

$$w_{\text{out}} = w_0[\lambda + \lambda z T + O(z + \lambda z^2 T + \lambda^3 \exp(-\lambda))]. \tag{3.9}$$

Again invoking the asymptotic properties of $N(t, \lambda)$, we deduce from (3.7) that with exponential accuracy $w(t) = w_{\text{out}}$ for $-T+2 \le t \le -3+\delta$. From this and the form of the substitution (3.6) we get the formula

$$\nu = (w_{\text{out}}/w_0)\exp(-zT) \tag{3.10}$$

for the leading multiplier of equation (3.5) (its remaining multipliers tend to zero as $\lambda \to \infty$). Equating ν to 1, we find from (3.10) that

$$z_{\text{cr}} = \lambda[\ln(\lambda \ln \lambda) + O(\ln\ln\lambda/\ln\lambda)]\exp(-\lambda). \tag{3.11}$$

We thus arrive at the following assertion.

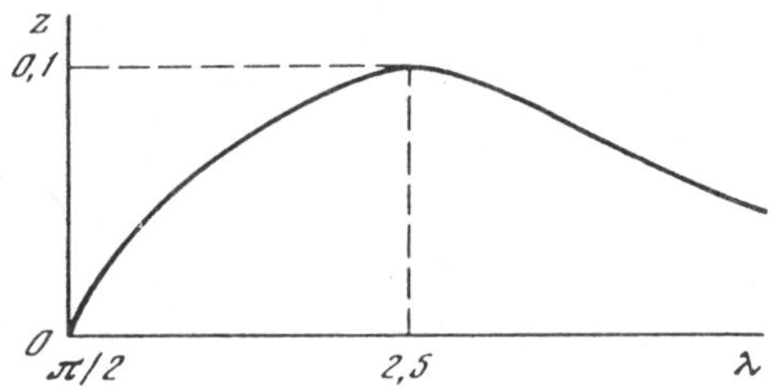

FIGURE 3.1

THEOREM 3.1. *For* $z < z_{cr}$ *equation* (3.5) *has exponentially increasing solutions.*

If $\lambda = \pi/2 + \varepsilon$, where $0 < \varepsilon \ll 1$, then the asymptotic expression for the neutral curve of equation (3.5) is constructed with the help of the local methods [40]. By combining the large-parameter and small-parameter methods with numerical integration, the neutral curve for arbitrary $\lambda > \pi/2$ is constructed in [12]. Its graph is shown in Figure 3.1.

3.3. The inferred dynamics. An h-oscillation of the system (3.1), (3.2) is defined to be a $2h$-periodic regime of it of the form

$$(N_*(t, \lambda), N_*(t-h, \lambda)). \tag{3.12}$$

It is easy to see that the system (3.1), (3.2) has an h-oscillation if and only if the equation

$$\dot{N} = D[N(t-h) - N] + \lambda[1 - N(t-1)]N \tag{3.13}$$

has a $2h$-periodic solution $N_*(t, \lambda)$.

THEOREM 3.2. *Suppose that* D *and* h *in* (3.13) *are arbitrarily fixed positive numbers. Then there is a sufficiently large positive number* λ_0 *such that for each* $\lambda \geq \lambda_0$ *equation* (3.13) *has a relaxation cycle with period asymptotically close to* $1 + h$.

The proof is relatively close to that of Theorem 2.1, and is thus omitted. Another reason is that the proof of a closely related but much more complicated result is considered in detail in the next chapter.

Equating to $2h$ the period of the cycle in Theorem 3.2, we get an equation that is obviously solvable, with h asymptotically close to 1. The latter means that under the conditions of Theorem 3.2 the system (3.1), (3.2) has h-oscillation (3.12) with period close to 2. Its average value is equal to $(2\lambda)^{-1}\exp\lambda$ with respect to order, and the smallest value of $N_*(t, \lambda)$ has only exponential smallness; that is, the operation of the system (3.1), (3.2) in an h-oscillation regime is advantageous from a biological point of view. However, under the conditions of Theorem 3.2 the h-oscillation (3.12) is unstable in the Lyapunov sense. We give a natural-science explanation of this fact.

For values of the diffusion D of order 1 we really have a single area that is conditionally divided into two. The homogeneous cycle (3.3) is stable according to Theorem 3.1, and the h-oscillation (3.12) is stable by assumption. We thus find that without special additional conditions the operation of this ecosystem (for example, a single-species ichthyocenosis) can be carried into the regime (3.12). It follows from properties of the h-oscillation that without marked losses for the population it is possible to remove from it a large number of individuals at certain moments of time separated by intervals of length at most two. We arrive at a peculiar kind of biological perpetual motion machine.

It turns out that an h-oscillation of the system (3.1), (3.2) is preserved for any decrease in the diffusion D. But its average value decreases, the minimum of the oscillations deepens, and the period grows.

A numerical analysis of the system (3.1), (3.2) for $2 \leq \lambda \leq 6$ (that is, for all values of this parameter of practical interest) enabled us to formulate the following statements.

For $D > z_{cr}/2$ the homogeneous cycle (3.3) is the unique attractor of the system (3.1), (3.2). For $z_{cr}/2 < D < D_*$ the attractor consists of two spatially heterogeneous cycles $(N_1(t, \lambda), N_2(t, \lambda))$ and $(N_2(t, \lambda), N_1(t, \lambda))$ branching off from the homogeneous cycle as D passes through $z_{cr}/2$. And as D passes through D_*, where D_*, as a function of λ, is superexponentially small, they attach to an h-oscillation, giving it their stability; that is, for $D < D_*$ the h-oscillation becomes the unique attractor. Further, there exists a $D_* < D_{max} < z_{cr}/2$ such that as D decreases from $z_{cr}/2$ to D_{max} the period of the spatially heterogeneous cycles is shortened, while the average value increases. As D decreases from D_{max} to D_*, their period begins to increase, and the average value decreases. The phase of the oscillations of the spatially heterogeneous cycles increases monotonically as D decreases from $z_{cr}/2$ to D_*.

We present some important numerical characteristics. The ratio $2D_{max}/z_{cr}$ increases monotonically to 0.1 for $2 \leq \lambda \leq 6$. For diffusion equal to D_{max} the period T_{max} of the oscillations decreases from values a little greater than three to values a little less than three as λ increases, the average value of the periodic function $N_{avg} = (N_1 + N_2)/2$ is a little greater than the quantity $(T_{max})^{-1} \exp \lambda$, and, finally, $\min N_{avg}$ is greater than 0.3.

In connection with what has been presented it becomes clear that in the practice of fishery management it is reasonable to connect ponds by suitably chosen channels—this will essentially increase the productivity (a good example of information leading directly to an increase in material goods).

Now a comment about the mathematical reason why a rigorous analysis of the dynamics of the system (3.1), (3.2) is difficult. It can be shown that for $D > z_{cr}/2$ the two-dimensional unstable manifold of the h-oscillation is simultaneously the stable manifold of the homogeneous cycle. This means that the system (3.1), (3.2) has a two-dimensional invariant torus that is exponentially stable, with the exponent of order λ. On the torus itself the motions are slow, and hence it is preserved under any decrease of D. Thus, the dynamic behavior of the system (3.1), (3.2) is periodic, but a more detailed treatment of it is made very complicated by the circumstance noted above.

4. The phenomenon of self-organization in a single-species biocenosis

4.1. Statement of the problem and heuristic considerations. The term "waves of life" was introduced in 1905 by S. S. Chetverikov, who pointed out its role in evolution. It was later observed [72] that oscillations in density are noticeably smaller in the central parts of a habitat area. This curious phenomenon was explained by the fact that the conditions for life in the center of the area best meet the requirements of the species.

The results of a numerical analysis of the boundary value problem (1.12) are presented below. In this connection we dwell in somewhat greater detail on the biological meaning of the diffusion coefficient D. It follows from (1.12) that D has the dimensions $\mathrm{km}^2/\mathrm{year}$. As we mentioned in §1, it is reasonable to assume that $D = D_0 \operatorname{meas} S$, where D_0 is a dimensionless coefficient and $S \subset \Omega$ is the region that an individual "masters" in the course of time h. A closely related meaning was given to the diffusion coefficient D in [65], where it was assumed that S is the disk of radius ρ, and $D_0 = 1/4\pi$.

It is convenient to assume that the domain Ω is unified, i.e.,

$$\operatorname{meas} \Omega = 1. \tag{4.1}$$

This can always be made to hold by the normalizations $x_1 \to kx_1$, $x_2 \to kx_2$. Here D is replaced by $k^{-2}D$. The case of a large habitat area is thus reduced to the case when the condition (4.1) holds, but with a small diffusion coefficient. Hence, under the condition (4.1) any values of the diffusion coefficient D in the boundary value problem (1.12) make biological sense.

In what follows we assume that h is taken as the unit of time in the boundary value problem (1.12). This means (see the beginning of § 2.1) that $h = 1$ in (1.12) and that the quantity $\lambda = rh$ stands in place of r.

Starting out mainly from biological considerations, we describe the conceivable stationary regimes of the boundary value problem (1.12) under the condition (4.1).

First, it is natural to assume that the number of different stationary regimes of the problem (1.12) increases without limit as the diffusion coefficient D decreases, and many of them bear a pronounced stochastic character. Both the first and the second statements are connected with the fact that the oscillations in the density of the local populations are weakly synchronized as D decreases, something mentioned briefly in §1.1.

Second, already for precritical values of D (that is, for $D > D_{cr}(\lambda)$), when the homogeneous cycle is still stable, spatially heterogeneous regimes with a complicated and at the same time nicely ordered spatial and temporal structure are formed from a condensation of trajectories. With reference to the first part of this statement it must be noted that in most cases a spatially homogeneous cycle loses stability strictly [21].* However, the purely biological explanation is more complete. With this goal we consider some local area. As follows from the second section, even for $\lambda > \pi/2$ not very large the homogeneous cycle has a high spike on a time interval of duration roughly equal to 1, after which the density becomes small for a relatively long time. Therefore, food is abundant during the corresponding time interval. It is understandable that at such times individuals from neighboring areas may migrate into the local area under consideration. If all this takes place rather consistently in all the local areas, then, on the one hand, the biological period of the oscillations (the time from one spike in the density to another, averaged over the local areas) decreases, and, on the other hand, the average value of the density increases. This mechanism can obviously function for D not too large, but it is not connected at all with the value of $D_{cr}(\lambda)$. It is also clear that these stabilizing factors exhibit themselves differently in the center of the habitat area (from which and into which it is convenient to migrate) and on its periphery.

* *Editor's note.* According to the first author, stability in the strict sese means here that the appearance of a stable cycle occurs under parameter values concomitant with a stable equilibrium.

It is expedient to call the stationary regimes described for the boundary value problem (1.12) self-organization regimes. Simultaneously with them, of course, there are spatially heterogeneous regimes of Hutchinson type whose biological periods are close to the period of the homogeneous cycle, their average values are close to 1, and the oscillations in the density are intensive in each of the local areas.

The spatial and temporal effects of the stationary regimes of the boundary value problem (1.12) must naturally exhibit themselves all the more intensely, the greater the fertility of the species, that is, the greater the reduced Malthusian coefficient of linear growth λ. However, essential difficulties in the numerical analysis of the problem (1.12) arise even for values of λ that are not very large. This is mainly connected with the fact that increasing λ leads to a decrease in the values of D for which the phenomena of interest to us are observed, and hence the spatially heterogeneous regimes become more and more "fine-grained". The last circumstance obliges us to take small steps in the space variables when we use numerical integration, and this leads to a significant increase in the number of nodes in the grid.

4.2. Spatially heterogeneous regimes in a square. Let us take Ω to be the unit square in the boundary value problem (1.12). We cover it by a uniform grid with step H and denote by $N_{k,m}(t)$ the values of the density N at the nodes of the grid. We approximate the boundary value problem (1.12) by the mappings

$$N_{k,m}(t+\tau) = [N_{k,m}(t) + \tau D \Delta_H N_{k,t}(t)] \times \exp\left[\lambda\tau\left(1 - \frac{N_{k,m}(t-1+\tau) + N_{k,m}(t-1)}{2}\right)\right], \tag{4.2}$$

$$k, m = 1, \dots, M-1, \qquad t = 0, \tau, 2\tau, \dots,$$

$$\begin{cases} N_{0,k} = N_{1,k}, \quad N_{M,k} = N_{M-1,k}, \quad k = 0, \dots, M, \\ N_{m,0} = N_{m,1}, \quad N_{m,M} = N_{m,M-1}, \quad m = 0, \dots, M, \end{cases} \tag{4.3}$$

where $M = 1/H$, and Δ_H is the usual difference approximation of the Laplace operator. We point out that the mappings (4.2), (4.3) have the same biological meaning as the boundary value problem (1.12). In §1.1 the special case (1.13), (1.14) of the mappings (4.2), (4.3) is proposed for an intuitive explanation of the process by which spatially heterogeneous structures arise in a homogeneous environment.

In our computations we set $H = 1/9$, $\tau = 0.01$, and $\lambda \le 3$, while we varied the diffusion coefficient D between broad limits. The initial conditions for $N_{k,m}(t)$ for $-1 \le t \le 0$ were taken in the form

$$N_{k,m}(t) = A_{k,m}\cos(\omega t + \alpha_{k,m}) + B_{k,m}. \tag{4.4}$$

We first note properties of the spatially heterogeneous stationary regimes of the boundary value problem (1.12) that were called Hutchinson regimes above. It turned out that they exist only for postcritical values of D. We describe the properties of two of them.

One, called a "wave traveling from corner to corner", has the following dynamics for $\lambda = 3$ and $D = 0.035$ (this value of the diffusion is just a little less than the critical value). The density peak arising at a corner of the square leads fairly quickly to a rise in the density in the whole square. Furthermore, in the course of time the density peak is displaced along the diagonal to the opposite corner. Then a low density is observed for a long time, after which the situation roughly repeats itself. Here the dynamic average is only a little greater than 1, and the biological period is roughly equal to 7.25; that is, it practically coincides with the Hutchinson period. We stress that the stationary regime under consideration is automatically not periodic, because the "measure" of the spatial heterogeneity changes essentially with the flow of time. It is likely that we are dealing with a "classical chaos" connected with the appearance of homoclinic structures.

Hutchinson regimes of a quite different type arise for $\lambda = 3$ and for smaller values of D. For example, for $D = 0.01$ a stationary regime is observed that is a wave traveling from the center of the square to its sides—a so-called leading center. Its biological characteristics, like those of the preceding regime, are not very nice.

We pass to the question of evolution of Hutchinson regimes as D decreases. For each of them the dynamic average $M[N_{\text{avg}}(t)]$, where

$$N_{\text{avg}}(t) = \iint_\Omega N(t, x_1, x_2)\, dx_1\, dx_2 \tag{4.5}$$

increases monotonically as D decreases. They also approach each other in their properties, since they are all transformed into a "diffusion chaos", when the period of the oscillations changes markedly upon passing from one local area to another, and this implies rather arbitrary change of the heights of the density peaks in the course of time.

In determining the self-organization regimes we chose the constants in (4.4) from the biological considerations already mentioned above: migration is always directed toward local areas in which an abundance of food has been observed. This reasonable consideration led to the discovery for $\lambda = 3$ of an important self-organization regime called a "vortex". Its evolution over time intervals equal to 0.25 is shown for $D = 0.035$ in Figure 4.1. We point out that the density spikes are maximal at the corners (of order 17.2–17.4; that is, they exceed the Hutchinson values), from which it is more difficult to migrate. In the four grid nodes closest to the center the density varies from 0.6 to 3; that is, the most favorable conditions for life are at the center of the area. The function (4.5) oscillates with amplitude of order 0.15 in a neighborhood of the value 2.9. The period of these oscillations is roughly equal to 0.75; this is connected with the symmetry of the domain.

The evolution of the vortex as the diffusion coefficient varies is of interest. It turns out that there is well-expressed vortex self-organization only as D decreases from 0.05 to 0.015.

We emphasize that for certain D a vortex thus coexists with a stable homogeneous cycle. There are three important regularities observed as the diffusion decreases. First, the biological period slowly increases as D decreases. For example, it is 2.85 for $D = 0.05$, while it becomes noticeably greater, equal to 3.25, for $D = 0.015$. Second, the dynamic average $M[N_{\text{avg}}(t)]$ decreases as D decreases (the situation is opposite to that for Hutchinson regimes). For example, for the respective values

$$D = 0.05;\ 0.035;\ 0.025;\ 0.015$$

we have

$$M[N_{\text{avg}}(t)] = 3.3;\ 2.9;\ 2.7;\ 2.5.$$

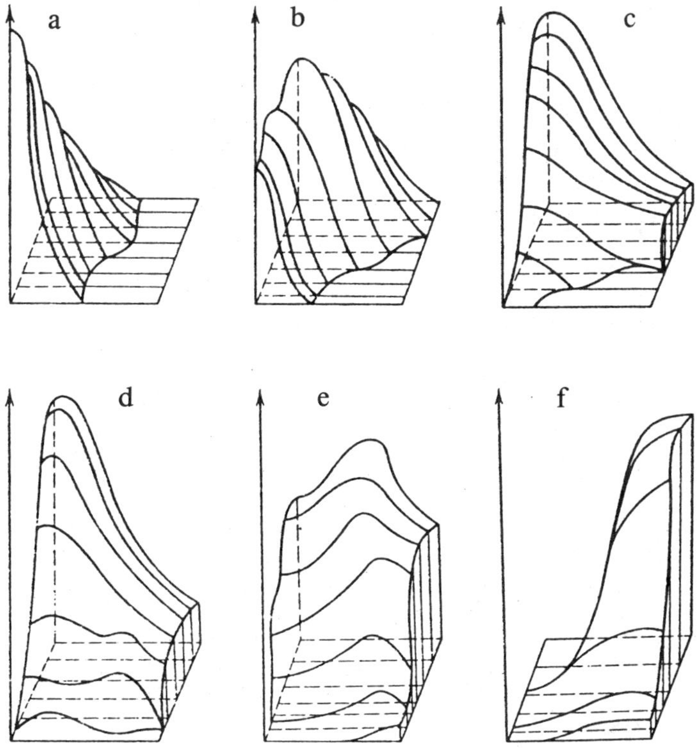

FIGURE 4.1

Third, as D decreases, the spatial and temporal structure of the vortex becomes less and less ordered. For this reason, self-organization regimes of a very different type arise.

We describe one such regime, which exists for $\lambda = 3$ only for postcritical values of D, and which we call a "stool". In determining it numerically we divide the square into four equal parts, in each of which initial conditions are specified that correspond to the vortex considered above. This means that a stool is a system of four coordinated vortices. For $D = 0.025$ its temporal development over time intervals equal to 0.5 is represented in Figure 4.2. We note in addition that the time during which the process goes through all its stages is roughly equal to 2.75, the maximal size of the density peaks is roughly equal to 14, and the integral density $N_{\text{avg}}(t)$ oscillates about 3.4 with amplitude of order 0.2. The stool exists and does not change qualitatively as D decreases from 0.03 to 0.01. However, while $M[N_{\text{avg}}(t)] = 3.4$ for $D = 0.025$, $M[N_{\text{avg}}(t)] = 3$ for $D = 0.01$. A uniform picture is thus observed: both in the case of a vortex and in the case of a stool a maximum of the temporal and spatial ordering happens at the moment of their origin, and they "collapse" smoothly as D gradually decreases.

Upon a significant decrease in D each of the two types of self-organization regimes described above is transformed into a diffusion chaos. Furthermore, they approach regimes of Hutchinson type, each of which, we recall, also passes into an analogous regime of operation as D decreases. We remark that in the diffusion chaos regime the quantity $M[N_{\text{avg}}(t)]$ falls only to a value close to 2 for $\lambda = 3$, and the biological period stays less than 3.75.

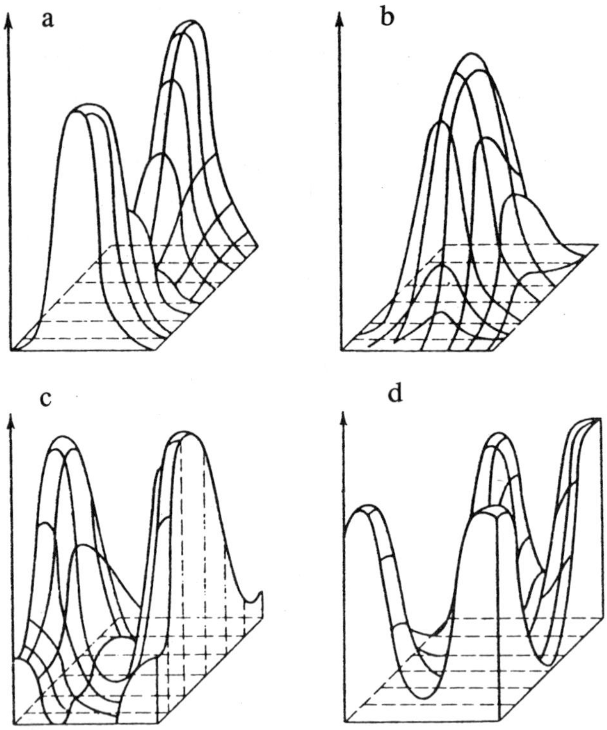

FIGURE 4.2

Finally, we mention the dynamic behavior of self-organization regimes that is connected with a decrease in λ. In full correspondence with the biological meaning, the biological characteristics singled out above become worse. The self-organization regime of vortex type is no longer observed for $\lambda < 2$ (the stool disappears earlier).

4.3. Spatially heterogeneous regimes in a triangle. To clear up the dependence of these regimes on the symmetry of the domain, we consider the case when Ω is a right triangle, $x_1, x_2 \geq 0$, $x_1 + x_2/2 \leq 1$. We cover it with a grid with steps $H_{x_1} = H$ with respect to x_1 and $H_{x_2} = 2H$ with respect to x_2. As a result, the boundary value problem (1.12) is replaced by the mappings (4.2), in which now $k + m \leq M - 1$, where $M = 1/H$ as before. Instead of (4.3) we now have

$$\begin{gathered} N_{k,M-k-1} = \tfrac{4}{5} N_{k-1,M-k-1} + \tfrac{1}{5} N_{k,M-k-2}, \qquad k = 2, \dots, M-3, \\ N_{1,M-2} = N_{1,M-3}, \qquad N_{M-2,1} = N_{M-3,2}, \end{gathered} \tag{4.6}$$

$$\begin{gathered} N_{0,m} = N_{1,m}, \qquad m = 0, \dots, M-1, \\ N_{k,0} = N_{k,1}, \qquad k = 0, \dots, M-1. \end{gathered} \tag{4.7}$$

Here the equalities (4.6) are approximations of the boundary conditions on the hypotenuse, and (4.7) are approximations on the legs. In the computation process we let $H = 0.1$, $\tau = 0.01$, and $\lambda = 3$, and we specified the initial conditions for $-1 \leq t \leq 0$ according to (4.4).

We dwell first on the spatially heterogeneous regimes of Hutchinson type. They correspond completely to what was found for the square: density waves traveling

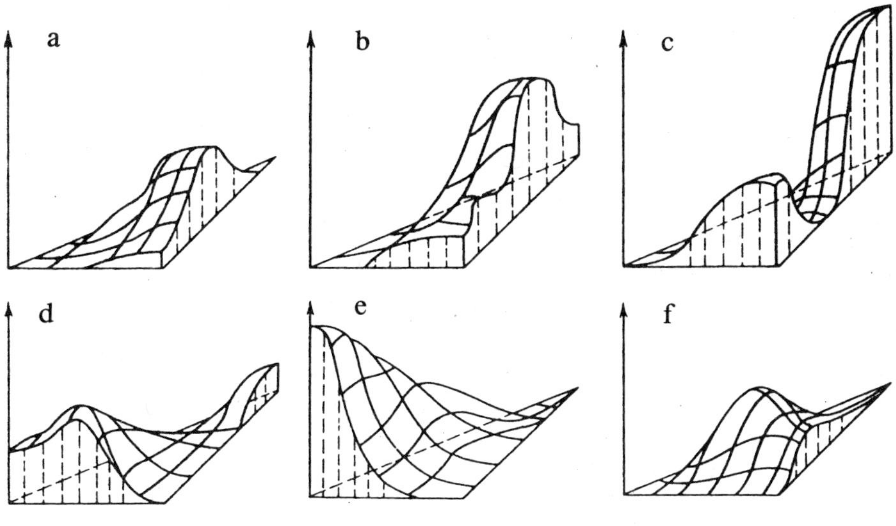

FIGURE 4.3

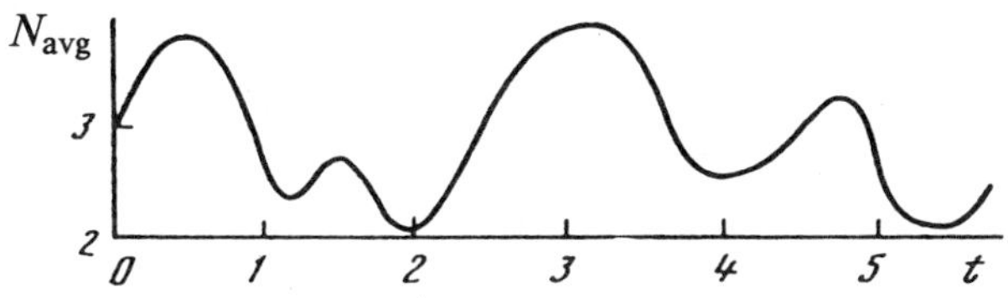

FIGURE 4.4

from one acute angle of the triangle to the other, and regimes of leading-center type. Their general properties are the same as in the case of the square, with each of them transformed into a diffusion chaos as D decreases significantly.

As in the case of a square, there is a self-organization regime of vortex type for relatively large D. For $D = 0.035$ its temporal development (over time intervals equal to 0.5) is shown in Figure 4.3. The graph of $N_{avg}(t)$ corresponding to this vortex is presented in Figure 4.4. It is quite clear that the oscillations of $N_{avg}(t)$ are not periodic, though they are close to being periodic (the periodicity was more strongly pronounced in the case of a square).

This self-organization regime is observed in a sufficiently broad range of variation of the diffusion coefficient, and, as in the preceding subsection, its biological period increases as D decreases, while the dynamic average density $M[N_{avg}(t)]$ decreases. For example, for $D = 0.04$ the biological period is close to 2.9, and for $D = 0.01$ it is close to 3.25. Further, $M[N_{avg}(t)]$ decreases from 3 to 2.8. It should be emphasized that as D decreases, the character of the behavior of the density waves at the angles of the triangle changes. First, the peaks become lower, which is a general pattern. Second, something which is specific to the triangle, the peaks of the density waves bifurcate upon passing the right angle and the acute angle adjacent to the unit leg: one of them "dies" at the angle, while the other moves on. Upon a significant decrease in D the vortex is regenerated into the diffusion chaos whose general properties are described in the preceding subsection.

Thus, we can regard it as established that both spatially heterogeneous regimes of Hutchinson type and self-organization regimes depend in a relatively weak way on the geometry of the domain.

4.4. Some comments. Summarizing the preceding results, we can assert that for large values of λ the spatially heterogeneous regimes of the boundary value problem (1.12) decompose into two strongly differing classes—regimes of Hutchinson type and self-organization regimes. Only for a significant decrease in the diffusion coefficient do the representatives of stationary regimes of these two types approach each other in their properties, passing into a diffusion chaos. We stress that it is individual stationary regimes we are discussing here, since for any decrease in D there arise more and more fine-grained self-organization regimes with excellent biological characteristics (cf. the vortex and stool constructed above).

The results we have obtained explain, in particular, the relatively small period of the oscillations in the density of species of mice in nature, a fact which (it seems to us) is connected with self-organization processes.

We dwell on the question of the character of periodic regimes of the system (3.1), (3.2) from new points of view. Comparing what was said in §3.3 and in this section, we conclude that for $D > D_{\text{max}}$ the spatially heterogeneous cycles of the system (3.1), (3.2) should be regarded as Hutchinson regimes. But for $D < D_{\text{max}}$ there is among the stationary regimes of the boundary value problem (1.12) no analogue of the cycles of the system (3.1), (3.2); that is, for these D they are specific to the biological situation described in §3.1.

If under the biological premises mentioned in §3.1 we consider three or more areas connected by relatively narrow passages, then the conclusion drawn in the preceding paragraph about the Hutchinson character of the stationary oscillations for $D > D_{\text{max}}$ is preserved [41] in general form (precritical oscillations are not observed for large values of λ). It is true that the periodicity of spatially heterogeneous regimes of the corresponding system of differential-difference equations is lost. But on the other hand, the optimal value of $M[N_{\text{avg}}(t)]$, which changes less as D deviates from D_{max}, becomes more stable.

The question of the dynamics of solutions of the boundary value problem (1.12) for large values of λ undoubtedly belongs to the theory of relaxation oscillations. However, the analytic difficulties arising are so great that we confined ourselves above to a description of numerical integration results that in our opinion are sufficiently complete. This has to do with the fact that the value $\lambda = 3$, and, all the more so, $\lambda = 6$, are not very small numbers; that is, the results obtained for these values can be extrapolated with great certainty to larger λ.

4.5. On one of the possible mechanisms for formation of species. The process of formation of species is the most complex and intricate question in the theory of evolution. It can be said conditionally that two extreme points of view on this problem have developed. Some researchers suppose that the formation of new species proceeds mainly due to internal factors—these are advocates of the so-called sympatric formation of species [19]. Others see the main cause in geographic isolation [56]. An intermediate point of view was expounded in [70], which stresses the important role of ecological factors in the mechanism of formation of species. It is this point of view to which the approach presented below is closest.

We consider a system

$$\frac{\partial N_1}{\partial t} = D\Delta N_1 + \frac{\lambda}{1+a}[1 + a(1 - N_2) - N_1(t-1, x)]N_1, \tag{4.8}$$

$$\frac{\partial N_2}{\partial t} = D\Delta N_2 + \frac{\lambda}{1+a}[1 + a(1 - N_1) - N_2(t-1, x)]N_2, \tag{4.9}$$

$$\left.\frac{\partial N_1}{\partial \nu}\right|_{\partial\Omega} = \left.\frac{\partial N_2}{\partial \nu}\right|_{\partial\Omega} = 0 \tag{4.10}$$

of differential-difference equations with diffusion that models a competitive struggle between two very close species, which shows in the equality of the diffusion coefficients, the Malthusian coefficients of linear growth, and the parameters characterizing the intensity of the competitive struggle.

It was shown above that for λ not small and D not large, complicated spatially heterogeneous regimes arise in the boundary value problem (1.12). This opens the possibility of realizing the following situation: two local populations (that is, two population waves separated in time) can enter into a competitive struggle upon encountering each other at some moment of time, and this leads mathematically to the boundary value problem (1.12) being replaced by the system (4.8)–(4.10). To better clarify matters we dwell briefly on the properties of stationary regimes of the boundary value problem (4.8)–(4.10). For example, self-organization regimes in which two vortices follow one after the other are possible for this problem, as well as more specific regimes. It is essential that on the most advantageous stationary regimes the dynamic average

$$\tfrac{1}{2}M[N_{1\,\text{avg}}(t) + N_{2\,\text{avg}}(t)] \tag{4.11}$$

is markedly greater than in the case of the boundary value problem (1.12).

Thus, we get that it is advantageous for a sufficiently fertile population to divide into at least two parts: this leads to an increase in the stability of the ecosystem, one measure of which can reasonably be taken to be the quantity (4.11).

5. The dynamics of the interaction of a predator with a strongly fertile prey

5.1. Statement of the problem. According to §1.2, to model interacting predator-prey populations we have the system of differential-difference equations

$$\dot{N}_1 = \frac{\lambda}{1+a}[1 + a(1 - N_2) - N_1(t-1)]N_1, \tag{5.1}$$

$$\dot{N}_2 = r[N_1 - N_2(t-h)]N_2. \tag{5.2}$$

Here N_1 and N_2 are the population densities of the prey and predator, respectively; the positive parameter a is the coefficient of the pressure of the predator on the prey, which leads to a reduction of the size of the prey population by a factor of $(1+a)$; h is the time span over which there is a replacement of generations in the population of the predator (the corresponding characteristic of the prey is taken as the unit of time), and r is the coefficient of linear growth of the predator; finally, λ is the Malthusian coefficient of linear growth of the prey, and is assumed to be large in what follows. As in §2, the latter means that the prey is strongly fertile.

If $N_2 \equiv 0$ (that is, there is no predator), then by making the normalization substitution $N_1 \to (1+a)N$ in (5.1) we arrive at the Hutchinson equation (2.1), whose relaxation cycle has the properties described in §2.

We give the following definition. A periodic regime of the system (5.1), (5.2) with positive components $N_1(t)$, $N_2(t)$ is called a k-cycle, $k = 0, 1, \dots$, if $N_1(t)$ varies according to the Hutchinson law in the sense that for this component the orders of the numerical characteristics in §2.2 of the cycle of equation (2.1) are analogous. We assume that $N_2(t)$ has one spike on time intervals exactly $k+1$ times greater than the period when $N_2(t) \geq 1$, the duration of the spike is close to h, and its height has order $\lambda^{-1}\exp\lambda$. The values of $N_2(t)$ are superexponentially small between spikes.

At first glance it seems likely that the k-cycles are natural attractors of the system (5.1), (5.2). However, as far back as 1980 the second author conjectured that for large values of λ chaotic oscillations are typical for such a system. At the basis of this conjecture is a simple biological consideration: as a consequence of the strong instability of the food base of the predator there must be a nonregular variation in its density, and hence also in the density of the prey. It is shown below that the situation is really much more complicated and intricate. Namely, it turns out that this biological statement is valid for large but not excessively large values of λ. In the limit there is degeneration of the strange attractor. These facts are of general theoretical interest.

5.2. Basic statements of the C-theory. We begin the exposition with the simplest case—conditions for the existence of 0-cycles. It follows from their very definition that the solutions of the system (5.1), (5.2) with the following positive initial conditions are to be considered.

Since $N_2(t)$ is superexponentially small during a time of order $\lambda^{-1}\exp\lambda$, it follows from §2 that at the end of this time interval $N_1(t)$ varies with high accuracy according to the Hutchinson law. It can therefore be assumed that for $-2 \leq t \leq -1$

$$N_1(t) = N(t, \lambda), \tag{5.3}$$

where

$$N(t, \lambda) = b\exp\left[\lambda(t+1) - \frac{b}{1+a}(\exp\lambda t - \exp(-\lambda)) + O(\exp(-2\lambda))\right],$$

and b is a positive constant whose role becomes clear later.

Now we consider initial conditions for the second coordinate of the solutions of the system (5.1), (5.2). It will be assumed that for $-1-h \leq t \leq -1$ the functions $N_2(t)$ increase monotonically, and

$$\underline{N}_2 \leq N_2(-1) \leq \overline{N}_2, \tag{5.4}$$

where

$$\underline{N}_2 = \exp[-\lambda^{-1}(B+\gamma_1)\exp\lambda], \tag{5.5}$$

$$\overline{N}_2 = \exp[-\lambda^{-1}(B+\gamma_2)\exp\lambda], \tag{5.6}$$

$$B = r(1+a)[1-\exp(-b/(1+a))], \tag{5.7}$$

and $\gamma_1 > \gamma_2$ are positive constants that will be chosen in a suitable way below.

To facilitate the perception of the material to be presented below, we describe somewhat coarsely the general character of behavior of the solutions of the system

(5.1), (5.2) with such initial conditions. For $-1 \leq t \leq 0$ the Hutchinson character of the variation of $N_1(t)$ is preserved, as follows from (5.7) and (5.5), (5.6). In fact, it is preserved up to an asymptotically small positive moment of time (that it has order λ^{-1} will be shown below) for which the second component of the solution $N_2(t)$ becomes of order 1. Beginning from this moment of time the variables in the system (5.1), (5.2) begin to interact strongly. However (and this is of essential significance), this time interval is exponentially small. But for just such a small time, $N_2(t)$ becomes a quantity of order $\lambda^{-1}\exp\lambda$, and $N_1(t)$ decreases from exponentially large values to quantities of order 1. As t increases further, the values of $N_1(t)$ become superexponentially small, while $N_2(t)$ is almost unchanged during a time of order h. For $t > h$ the values of $N_2(t)$ also become superexponentially small. For these t the difference between $N_1(t)$ and $N_2(t)$ manifests itself in the following. The variable $N_1(t)$ begins to change in good agreement with the Hutchinson law, while $N_2(t)$ is practically unchanged up to the next spike in the prey density.

We let

$$\gamma = -\lambda\exp(-\lambda)\cdot\ln N_2(-1) - B, \tag{5.8}$$

and we introduce the function

$$F(\gamma) = r\lambda\int_0^{t_1} N_1(s)\,ds\cdot\exp(-\lambda), \tag{5.9}$$

where t_1 is an asymptotically small moment of time such that

$$N_1(t_1) = 1. \tag{5.10}$$

LEMMA 5.1. *Suppose that*

$$\gamma > r(1+a)\exp(-b/(1+a)). \tag{5.11}$$

Then

$$F(\gamma) = r(1+a)\exp(-b/(1+a)) + o(1). \tag{5.12}$$

PROOF. It follows from (5.4) and (5.5)–(5.8) that under the condition (5.11) the predator density remains superexponentially small also for $0 \leq t \leq t_1$. From this and (5.9) we conclude that the asymptotic equality (5.12) holds.

LEMMA 5.2. *Suppose that*

$$0 < \gamma < r(1+a)\exp(-b/(1+a)). \tag{5.13}$$

Then

$$F(\gamma) = \gamma + o(1). \tag{5.14}$$

PROOF. We introduce the notation

$$J(t) = \frac{a\lambda}{1+a}\int_0^t N_2(s)\,ds. \tag{5.15}$$

It follows from (5.1) and (5.10) that with high accuracy

$$J(t_1) = \ln N(t_1, \lambda). \tag{5.16}$$

Maximizing the right-hand side of (5.16) with respect to t_1, we see that

$$J(t_1) \leq \lambda. \tag{5.17}$$

It follows from (5.2) that for $0 \le t \le t_1$

$$\dot{N}_2 \le r N_1(t) N_2(t) . \tag{5.18}$$

Further, according to (5.1) and (5.3),

$$N_1(t) \le (1+a)\exp\lambda , \qquad 0 \le t \le t_1 . \tag{5.19}$$

Using (5.17) and (5.19), we deduce from the differential inequality (5.18) that

$$N_2(t_1) \le \frac{2r(1+a)^2}{a} \exp\lambda . \tag{5.20}$$

We proceed to the proof of the asymptotic equality (5.14). Assume first that

$$F(\gamma) > \gamma \tag{5.21}$$

for some γ satisfying (5.13). We get from obvious properties of the system (5.1), (5.2) that the predator density is superexponentially large at the time t_1. This contradicts the estimate (5.20); that is, (5.21) is impossible. It remains to consider the case when

$$F(\gamma) < \gamma . \tag{5.22}$$

Under the condition (5.22) the predator density is superexponentially small for $0 \le t \le t_1$. But then the asymptotic equality (5.12) holds. Comparing it with (5.22) and the right-hand inequality in (5.13), we see that we have a contradiction. The proof is complete.

We assume that under the conditions of Lemma 5.2 the quantity $N_2(t_1)$ is asymptotically small. The asymptotic equality (5.12) then follows from (5.10). We arrive at a contradiction to the right-hand inequality in (5.13).

Accordingly, under the conditions (5.13) $N_2(t_1)$ is certainly not small. For technical reasons it is convenient to assume that $N_2(t_1) \ge 1$. We emphasize once more that the choice of 1 is due to considerations of convenience.

Denote by t_0 a moment of time such that

$$N_2(t_0) = 1 , \qquad 0 < t_0 \le t_1 . \tag{5.23}$$

LEMMA 5.3. *The following asymptotic equality holds*:

$$t_0 = \left\{ \ln\left[1 - \frac{1+a}{b}\ln(1-y)\right] + o(1) \right\} / \lambda , \tag{5.24}$$

where

$$y = \gamma/[r(1+a)\exp(-b/(1+a))] . \tag{5.25}$$

PROOF. We remark that t_0 is a moment of time of superexponential growth of $N_2(t)$. Therefore, the integral of $N_2(t)$ over the interval $0 \le t \le t_0$ is exponentially small. The formula (5.3) is thereby preserved on the whole for the indicated t. It is also essential that in view of (5.20) the integral of $N_1(t)$ over the interval $t_0 \le t \le t_1$ does not exceed λ with respect to order. Therefore, to find t_0 we can use (5.9) and the asymptotic equality (5.14). In this way we arrive at (5.24).

Let

$$x(t) = \int_{t_0}^{t} N_1(s)\, ds . \tag{5.26}$$

It follows from obvious properties of the system (5.1), (5.2) that with superexponential accuracy

$$\frac{\ddot{x}}{\dot{x}} = \lambda\left[1 - \frac{N_1(t-1)}{1+a}\right] - \frac{a\lambda}{1+a}\exp rx\,. \tag{5.27}$$

Multiplying by $\dot{x}$ and then integrating, we pass from (5.27) to the integro-differential equation

$$\begin{aligned}\dot{x} = \dot{x}(t_0) + \lambda\left[1 - \frac{N_1(t-1)}{1+a}\right]x + \frac{\lambda}{1+a}\int_{t_0}^{t}\dot{N}_1(s-1)x(s)\,ds \\ + \frac{a\lambda}{r(1+a)}(1-\exp rx)\,.\end{aligned} \tag{5.28}$$

In place of (5.28) we first consider the essentially simpler equation

$$\dot{x} = \dot{x}(t_0)[1 - \exp r(x - \tilde{x})]\,, \tag{5.29}$$

where

$$\tilde{x} = \frac{\lambda}{r} - \frac{\ln\lambda}{r} + \frac{\ln A}{r}\,, \qquad A = \frac{r(1+a)}{a}N_1(t_0)\exp(-\lambda)\,. \tag{5.30}$$

For the time being we work with the model equation (5.29) as if it were the exact equation.

By (5.29) and (5.30),

$$J(t_1) = J(t_0) + r[\dot{x}(t_0)(t_1 - t_0) - x(t_1)]\,. \tag{5.31}$$

According to (5.2),

$$J(t_0) = [1 + o(1)]\frac{a\lambda}{r(1+a)N_1(t_0)}\,, \tag{5.32}$$

which, by (5.10) and (5.29)–(5.31), implies that

$$t_1 - t_0 = \frac{2\lambda - \ln\lambda}{rN_1(t_0)} + O(\exp(-\lambda))\,. \tag{5.33}$$

LEMMA 5.4. *Let $x(t)$, $x(t_0) = 0$, be a solution of the integro-differential equation* (5.28). *Then the asymptotic equality* (5.33) *holds for the quantity $t_1 - t_0$, and*

$$x(t_1) = \tilde{x} + O(\lambda^2\exp(-\lambda))\,. \tag{5.34}$$

We shall not dwell on the rather simple proof, which is based on an analysis of the equation for the difference between the corresponding solutions of (5.28) and (5.29).

From (5.34) we get an asymptotic equality that is important for what follows:

$$N_2(t_1) = [1 + o(1)]\lambda^{-1}A_{np}\exp\lambda\,, \tag{5.35}$$

where, by (5.24) and (5.25),

$$A_{np} = \frac{r(1+a)}{a}(1-y)[b - (1+a)\ln(1-y)]\exp\left(-\frac{b}{1+a}\right)\,. \tag{5.36}$$

For $t \geq 3h$ we deduce from (5.1) and (5.2) the approximation formula

$$N_2(t) = N_2(t_1)\exp[-rhN_2(t_1)]\,, \tag{5.37}$$

which is preserved up to the next spike in the prey density.

It follows from (5.35)–(5.37) that an essential role in the question of the existence of 0-cycles must be played by the mapping

$$y \to F(y), \tag{5.38}$$

where

$$F(y) = 1 - \exp\frac{b}{1+a} + rh\frac{b}{a}(1-y) - \Delta(1-y)\ln(1-y),$$

and

$$\Delta = rh(1+a)/a. \tag{5.39}$$

We remark that by means of the substitution

$$z = (1-y)\exp(-b/(1+a)) \tag{5.40}$$

the mapping (5.38) is transformed into the very simple form

$$z \to 1 + \Delta z \ln z. \tag{5.41}$$

According to (5.25) and (5.40), in (5.41)

$$0 < z < \exp(-b/(1+a)). \tag{5.42}$$

Let

$$1 < \Delta < e. \tag{5.43}$$

It then follows from the structure of the mapping (5.41) that for suitably small b it transforms some interval

$$0 < z_1 \le z \le z_2 < 1 \tag{5.44}$$

into itself. It can thus be expected that the system (5.1), (5.2) has a 0-cycle under the condition (5.43).

Denote by S_1 the set of continuous functions of the form (5.3) such that the remainder in the formula for $N(t, \lambda)$ can be estimated in modulus by $M_1 \exp(-2\lambda)$, where the positive constant M_1 is suitably large. Denote by S_2 the set of monotonically increasing continuous functions on $-1-h \le t \le -1$ such that the inequalities (5.4) hold at $t = -1$, where the constants γ_1 and γ_2 in (5.5) and (5.6), upon recomputation with the help of (5.25), pass into constants y_1 and y_2, $0 < y_2 < y_1 < 1$, such that the mapping (5.38) takes the segment $[y_2, y_1]$ into itself.

Our next theorem follows immediately from the above constructions.

THEOREM 5.1. *Assume the inequalities* (5.43). *Then there is a sufficiently large positive number* λ_0 *such that for* $\lambda \ge \lambda_0$ *the first-return operator along trajectories of the system* (5.1), (5.2) *is defined on the closed convex subset* $S_1 \times S_2$ *of* $C(-2, -1) \times C(-1-h, -1)$ *and transforms it into itself.*

To prove the C^1-theorem we need a more subtle connection between the fixed point y_0 of the mapping (5.38) and the properties of the cycle which exists under the conditions of Theorem 5.1. We begin with the case when y_0 is exponentially stable, and it will have this property only when

$$1 < \Delta < 2.25\cdots. \tag{5.45}$$

THEOREM 5.2. *Assume the inequalities* (5.45). *Then there is a sufficiently large positive number* λ_0 *such that for* $\lambda \geq \lambda_0$ *the first-return operator along trajectories of the system* (5.1), (5.2) *transforms the closed convex set* $S_1 \times S_2$ *in Theorem* 5.1 *into itself, except that now the constants* y_1 *and* y_2 *in the definition of* S_2 *are such that*

$$y_1 = y_0 + \delta(\lambda), \quad y_2 = y_0 - \delta(\lambda), \tag{5.46}$$

where the positive function $\delta(\lambda) \to 0$ *as* $\lambda \to \infty$.

THEOREM 5.3. *Suppose that*

$$2.25\cdots < \Delta < \infty. \tag{5.47}$$

Then there exists a sufficiently large positive number λ_0 *such that for* $\lambda \geq \lambda_0$ *the first-return operator along trajectories of the system* (5.1), (5.2) *is defined on the set* $S_1 \times S_2$ *in Theorem* 5.2, *is contracting with respect to the variable* N_1, *and is expanding with respect to* N_2.

The proofs of these assertions also follow from the above constructions.

If the mapping (5.41) has an exponentially stable or exponentially unstable cycle

$$z_1, \ldots, z_m, \tag{5.48}$$

then Theorems 5.2 and 5.3 can be modified without difficulty to correspond to this situation.

We return to the inequalities (5.43) in order to better explain their role.

For $\Delta < 1$ the equilibrium state $z = 1$ of the mapping (5.41) is exponentially stable, and hence in view of the inequalities in (5.42) any decrease in b takes us outside the bounds of applicability of Theorems 5.1 and 5.2. The situation arising when Δ decreases and passes through 1 is nicely explained by the mapping (5.38). We can easily see that for

$$\Delta = \Delta(b) = \left(\exp\frac{b}{1+a} - 1\right) \bigg/ \frac{b}{1+a}$$

its zero equilibrium state is exponentially stable. For $1 < \Delta < \Delta(b)$ the function $F(y)$ is negative for small values of y. It is essential that for the indicated Δ each trajectory of the mapping (5.38) with an arbitrary initial condition $0 < y < 1$ such that $F(y) > 0$, drops to negative values of y. This means that there is a transition to stationary regimes that are biologically more advantageous for the ecosystem under consideration.

We now turn to the case $\Delta > e$. For definiteness it will be assumed that

$$e < \Delta < 2e. \tag{5.49}$$

Under the condition (5.49) the following circumstance is fundamental: almost every trajectory of the mapping (5.38) with initial condition y such that $0 < F(y) < 1$ drops to values of y for which $F(y) > 1$. By (5.49), for small values of b it is not a serious restriction to assume that in fact

$$\exp\left(\frac{b}{1+a}\right) < F(y) < 1 + \exp\left(\frac{b}{1+a}\right). \tag{5.50}$$

The inequalities (5.50) clarify the situation. Indeed, suppose that the quantity $\tilde{y}$ given on the input satisfies the inequalities (5.50). Then upon consideration of the next spike in the prey density (by (5.50), the predator hardly reacts to the first spike in it) the role of $\tilde{y}$ will be filled by the quantity

$$y = \tilde{y} - \exp\left(\frac{b}{1+a}\right). \tag{5.51}$$

We take this into account in the mapping $\tilde{y} \to F(\tilde{y})$, in which we then make the substitution (5.40). As a result, we get the mapping $z \to 2 + \Delta z \ln z$, which works for $\underline{z} < z < \overline{z}$, where $\underline{z}$ and $\overline{z}$ are the roots of the equation $z \ln z = -1/\Delta$. Thus, under the condition (5.49) the mapping

$$z \to \Phi_1(z) \tag{5.52}$$

is responsible for the dynamics, where

$$\Phi_1(z) = \begin{cases} 1 + \Delta z \ln z, & z \notin [\underline{z}, \overline{z}], \\ 2 + \Delta z \ln z, & z \in [\underline{z}, \overline{z}]. \end{cases}$$

The mapping (5.52) allows us to analyze the question of the existence of 1-cycles in a relatively simple way. Our construction permits us to detect cycles with large indices only for $\Delta > 2e$. Here there arises a mapping $z \to \Phi_k(z)$, depending on the index of the k-cycle, that is analogous to (5.52) but has a more complicated structure. It is clear how to construct it, and hence there is no sense in dwelling on this in greater detail.

5.3. Proof of the C^1-theorem. We linearize the system (5.1), (5.2) on the 0-cycle referred to in Theorem 5.2 or Theorem 5.3, and then we make normalization substitutions analogous to (2.13). We arrive finally at the linear system

$$\dot{v} = -\frac{\lambda}{1+a} N_1(t-1)v(t-1) - \frac{a\lambda}{1+a} N_2(t)w, \tag{5.53}$$

$$\dot{w} = rN_1(t)v - rN_2(t-h)w(t-h). \tag{5.54}$$

Arguing as in §2.4 (true, the estimate analogous to (2.20) is obtained in a slightly more complicated way: its derivation is based on a lemma on integral inequalities in systems with delay, the simplest variant of which is given in, for example, [34] and [50]), we see that with increasing time the solutions of the system (5.53), (5.54) collapse during a relatively short time to a two-dimensional family. On the algorithmic level this means the following (cf. the exposition in §2.3). Let

$$v(t) \in C(-2, -1), \qquad w(t) \in C(-1-h, -1).$$

Then already for $t = 3h$ the Cauchy solving operator of the system (5.53), (5.54) can be approximated with superexponential accuracy by a two-dimensional matrix applied to the vector (v_0, w_0), where v_0 and w_0 are the values of the initial conditions at $t = -1$. There is another simplifying circumstance. The system (5.53), (5.54) has a periodic solution with coordinates

$$\frac{1}{\lambda}\frac{\dot{N}_1(t)}{N_1(t)}, \quad \frac{1}{\lambda}\frac{\dot{N}_2(t)}{N_2(t)}. \tag{5.55}$$

For $t = -1$ we have with high accuracy that $v_0 = 1$ and $w_0 = 0$ for the quantities in (5.55). With superexponential accuracy the same vector $(1, 0)$ is obtained also for $t = 3h$. Therefore, the matrix of interest to us has the form

$$\begin{pmatrix} 1 & b_1(\lambda) \\ 0 & b_2(\lambda) \end{pmatrix}. \tag{5.56}$$

It follows from (5.56) that the problem reduces to the asymptotic integration of the system (5.53), (5.54) with the initial conditions

$$v(t) \equiv 0, \quad w(t) \equiv 1, \qquad t \leq -1. \tag{5.57}$$

It is sufficiently clear that for $-1 \leq t \leq t_0$ the second identity in (5.57) holds with high accuracy. This implies that, by (5.53) and (5.54), $\dot{w}(t_0)$ satisfies the formula

$$\dot{w}(t_0) = -\frac{ar\lambda}{1+a} N_1(t_0) \int_0^{t_0} N_2(s)\, ds. \tag{5.58}$$

From (5.15), (5.32), and (5.58) we deduce that

$$\dot{w}(t_0) = -[1 + o(1)]a\lambda/(1+a). \tag{5.59}$$

Therefore, for $t_0 \leq t \leq t_1$ we study the Cauchy problem

$$\ddot{w} - \frac{\dot{N}_1(t)}{N_1(t)}\dot{w} + \frac{ar\lambda}{1+a} N_1(t) N_2(t) w = 0, \tag{5.60}$$

$$w(t_0) = 1, \qquad \dot{w}(t_0) = -a\lambda/(1+a). \tag{5.61}$$

Equation (5.2) tells us that on the necessary time interval we have, with superexponential accuracy,

$$\dot{N}_2 = rN_1(t)N_2(t). \tag{5.62}$$

Taking (5.61) and (5.62) into account, we integrate (5.60) over $[t_0, t]$ and write the result in the form

$$\begin{aligned} \dot{w} &= \lambda \left[-\frac{a}{1+a} N_2(t) + 1 - \frac{N_1(t-1)}{1+a} \right] w \\ &\quad + \frac{\lambda}{1+a} \left[\int_{t_0}^{t} \dot{N}_1(s-1) w(s)\, ds - 1 - a + N_1(t_0 - 1) \right]. \end{aligned} \tag{5.63}$$

From the subsequent constructions it will follow that the integro-differential equation (5.63) can be replaced by the ordinary equation

$$\dot{w} = -\frac{a\lambda}{1+a} N_2(t) w - \frac{\lambda}{1+a}[1 + a - N_1(t_0 - 1)]. \tag{5.64}$$

From (5.64) and the first equality in (5.61) we get

$$w(t_1) = \exp(-J(t_1, t_0)) - \frac{\lambda}{1+a}[1 + a - N_1(t_0 - 1)] \int_{t_0}^{t_1} \exp(-J(t_1, \tau))\, d\tau, \tag{5.65}$$

where (cf. (5.15))

$$J(t_1, \tau) = \frac{a\lambda}{1+a} \int_\tau^{t_1} N_2(s)\, ds. \tag{5.66}$$

Further progress involves the equation

$$\dot{x} = \dot{x}(t_0)[1 - \exp r(x - x(t_1))], \tag{5.67}$$

which, by Lemma 5.4, does not differ essentially from (5.29) nor from (5.28). From (5.30), (5.34), and (5.67) we deduce that with acceptable accuracy

$$J(t_1, \tau) = r[x(\tau) - x(t_1) + \dot{x}(t_0)(t_1 - \tau)], \tag{5.68}$$

and hence (cf. (5.31))

$$J(t_1, t_0) = r[\dot{x}(t_0)(t_1 - t_0) - x(t_1)]. \tag{5.69}$$

It follows from (5.30), (5.33), (5.34), and (5.69) that the first term on the right-hand side of (5.65) has order $\exp(-\lambda)$, and this order of smallness (as will be clear from what follows) permits it to be ignored.

To analyze the question of the order of smallness of the second term on the right-hand side of (5.65) we take into account in (5.68) the explicit formula

$$x(t) = \dot{x}(t_0)(t - t_0) - \frac{1}{r}\ln[1 - \exp(-rx(t_1)) + \exp(r(\dot{x}(t_0)(t - t_0) - x(t_1)))] \tag{5.70}$$

for the solution $x(t)$, $x(t_0) = 0$, of equation (5.67). As a result we see that

$$\begin{aligned}\exp[-J(t_1, \tau)] = {} & \exp[-J(t_1, t_0)] - \exp[-r\dot{x}(t_0)(t_1 - t_0)] \\ & + \exp[-r\dot{x}(t_0)(t_1 - \tau)].\end{aligned} \tag{5.71}$$

The formula (5.71) enables us to get the important relation

$$\lambda^{-1} w(t_1)\exp\lambda = \frac{1 + \ln z_0}{r(1 + a)z_0 \ln z_0} + o(1), \tag{5.72}$$

where, according to (5.40),

$$z_0 = (1 - y_0)\exp(-b/(1 + a)). \tag{5.73}$$

We emphasize that (5.72) is preserved also for the corresponding solution of (5.63). We omit the trivial proof, which involves the use of a theorem on differential inequalities.

In view of the sharp decrease in $N_1(t)$ and the structure of the system (5.53), (5.54) we have

$$w(t) \equiv w(t_1), \qquad t_1 \le t \le h + t_0, \tag{5.74}$$

with high accuracy. For $h + t_0 \le t \le h + t_1$ it follows from (5.53), (5.54), and (5.74) that

$$w(t) = w(t_1) - r\int_{t_0}^{t-h} N_2(s)w(s)\,ds + o(1). \tag{5.75}$$

From (5.10) and (5.60), (5.61) we deduce that

$$-r\int_{t_0}^{t_1} N_2(s)w(s)\,ds = \frac{1 + a}{a\lambda}\dot{w}(t_1). \tag{5.76}$$

It is important that, by the asymptotic equality (5.72),

$$\lambda^{-1}\dot{w}(t_1) = o(1). \tag{5.77}$$

Taking (5.76) and (5.77) into account in (5.75), we conclude that

$$w(h + t_1) = o(1). \tag{5.78}$$

According to the preceding subsection,

$$N_2(t) = [1 + o(1)]\lambda^{-1}A_{np}\exp\lambda, \qquad t_1 \le t \le h + t_0. \tag{5.79}$$

It follows from (5.53), (5.54) and (5.74), (5.78), (5.79) that for $h+t_1 \le t \le 2h+t_0$

$$w(t) = -(t - h - t_1)rN_2(t_1)w(t_1) + o(1). \tag{5.80}$$

We pass to the interval $2h + t_0 \le t \le 2h + t_1$. According to (5.53) and (5.54), on this time interval

$$w(t) = w(2h + t_0) - r \int_{h+t_0}^{t-h} N_2(s) w(s)\, ds + o(1). \tag{5.81}$$

Taking (5.75) into account in (5.81), we get

$$\begin{aligned} w(2h + t_1) = w(2h + t_0) - r w(t_1) \int_{h+t_0}^{h+t_1} N_2(s)\, ds \\ + r^2 \int_{h+t_0}^{h+t_1} N_2(\sigma) \int_{t_0}^{s-h} N_2(\sigma) w(\sigma)\, d\sigma\, ds + o(1). \end{aligned} \tag{5.82}$$

The integral of $N_2(t)$ over the interval $[h+t_0,\, h+t_1]$ is finite. Therefore, the second term on the right-hand side of (5.82) is small in view of the smallness of $w(t_1)$. To estimate the third term we note that, by (5.64),

$$\left| \int_{t_0}^{s-h} N_2(\sigma) w(\sigma)\, d\sigma \right| \le \frac{c}{\lambda},$$

where the positive constant c does not depend on λ. We recall also that for $h+t_0 \le t \le h + t_1$ the values of $N_2(t)$ are equal to $\lambda^{-1} \exp \lambda$ with respect to order, and the asymptotic representation (5.33) holds for the difference $t_1 - t_0$. We conclude that

$$w(2h + t_1) = w(2h + t_0) + o(1). \tag{5.83}$$

It follows from (5.53), (5.54), and (5.83) that

$$w(3h) = w(2h + t_0) - r \int_{h+t_1}^{2h} N_2(s) w(s)\, ds + o(1). \tag{5.84}$$

On the right-hand side we take into account (5.80). Using (5.79) in passing, along with the features of equation (5.2), we deduce that

$$w(3h) = w(2h + t_0) - w(h + t_1) \exp\left(-r \int_{t_0}^{t_1} N_2(s)\, ds \right) + o(1). \tag{5.85}$$

We derive the final result from (5.78)–(5.80) and (5.85):

$$w(3h) = -rhA_{np} \lambda^{-1} w(t_1) \exp \lambda + o(1). \tag{5.86}$$

Here we take (5.36) and (5.72) into account, obtaining

$$b_2(\lambda) = \Delta(1 + \ln z_0) + o(1). \tag{5.87}$$

This equality enables us to formulate the following assertion.

THEOREM 5.4. *Under the conditions of Theorems* 5.2 *and* 5.3, *countably many multipliers of the system* (5.53), (5.54) *tend to zero as* $\lambda \to \infty$, *one of them is equal to* 1, *and the asymptotic equality* (5.87) *holds for another one of them.*

An assertion analogous to Theorem 5.4 is valid in the question of stability of k-cycles for $1 \le k \le k_0$. A suitable modification of Theorem 5.4 holds also in the question of stability of the cycles corresponding to the cycles (5.48) of the mapping (5.41).

We make one more remark. An insignificant modification of the proof of Theorem 5.4 allows us to see that on the set $S_1 \times S_2$ introduced in Theorem 5.1 the first-return operator along trajectories of the system (5.1), (5.2) is close to the mapping (5.41) in the C^1-metric. This additional information enables us to link the dynamical properties of the system (5.1), (5.2) more closely with the mapping (5.41). For example, with the help of arguments in [30] it is possible to establish that the original system (5.1), (5.2) has countably many dichotomous cycles in the case when a cycle of the form (5.48) for the mapping (5.41) has a structurally stable homoclinic trajectory.

5.4. On the limits of applicability of the asymptotic theory, and on global dynamics. We begin with an exposition of results of numerical integration of the system (5.1), (5.2). In the computation process the parameters a and h are fixed in different ways, after which the parameter r is gradually increased for a series of sufficiently large λ. It has become clear that for relatively small r the attractor of solutions with randomly chosen positive initial conditions is a 0-cycle; that is, the predator reacts to each spike in the prey density. As r passes through a certain value $R_0(\lambda, a, h)$ there is a period-doubling bifurcation of it that manifests itself mainly in adjacent peaks of the predator density begining to differ somewhat in height. As r is further increased slightly, there is a period-quadrupling bifurcation, and so on. As a result, stochastic oscillations are established that transform into periodic oscillations upon further increase of r, then chaos again ensues, and so on. However, as λ increases, the width (with respect to r) of the zones in which a stable cycle arises from the chaos decreases sharply.

It also turned out that, at least for small a, a stable 1-cycle, 2-cycle, etc., arises from the condensation of trajectories as λ increases. Each of them has a period-doubling bifurcation, and so on, as r passes through certain values $R_k(\lambda, a, h)$, where

$$R_0(\lambda, a, h) < R_1(\lambda, a, h) < R_2(\lambda, a, h) < \cdots .$$

One more property: the functions $R_k(\lambda, a, h)$ are monotonically decreasing in λ and h, and monotonically increasing in a. Further, for small values of a the functions $R_k(\lambda, a, h)$ with small indices noticeably approach zero as λ increases. However, their decrease becomes insignificant as a increases. Finally, as λ increases, it becomes more difficult to catch stable k-cycles with large indices, which indicates that their domains of attraction are decreasing.

Intuitive biological arguments expressed by the second author in 1980 served as a theoretical basis for the numerical integration results described.

If the food base for the predator becomes more and more unstable (the parameter λ increases), then it cannot effectively control the situation in general. This implies, in particular, that the prey density varies on significant time intervals in good agreement with the Hutchinson law. We automatically get that not only is the predator density small between peaks of the prey density, but also its real coefficient of linear growth $rN_1(t)$ is small there. Hence, it is appropriate to say that stress is observed in the predator population in the corresponding time intervals—its reproductive capacity is low in the course of several generations. And in the general case the predator reacts to high spikes in the prey density by population explosions, a fact which serves as the main cause of chaotic change in time of both its density and the prey density. Here the biological meaning of a strange attractor is that the latter is more advantageous than the regular processes: the "troughs" of $N_1(t)$ and $N_2(t)$ are much more moderate, and hence the ecosystem under consideration is more stable.

It is quite clear that our theoretical results do not confirm the biological considerations presented and do not correspond to the numerical experiments described above. The reason for this is not trivial. We proceed to an explanation of it.

THEOREM 5.5. *Suppose that in the system* (5.1), (5.2)

$$a = a_0/\lambda, \quad r = r_0/\lambda, \tag{5.88}$$

where the positive constants a_0 *and* r_0 *do not depend on* λ. *Then there is a sufficiently large positive number* λ_0 *such that the assertions of Theorems* 5.1–5.4 *hold for* $\lambda \geq \lambda_0$, *but now the role of* Δ *in the mappings* (5.41), (5.52), *etc., is filled by the quantity* $r_0 h/a_0$.

The assertion of Theorem 5.5 is preserved if the parameters a and r in the system (5.1), (5.2) are arbitrary fading functions of λ, but here the ratio r/a has a finite limit as $\lambda \to \infty$, and, moreover,

$$\lim_{\lambda \to \infty} a \exp \gamma\lambda = \lim_{\lambda \to \infty} r \exp \gamma\lambda = \infty$$

for some $0 < \gamma < 1$; that is, the functions a and r are not too rapidly decreasing.

THEOREM 5.6. *Suppose that in the system* (5.1), (5.2) *the degenerate case holds in which the coefficient* a *of the pressure of the predator on the prey is* 0. *Then there exist a sufficiently large positive number* λ_0 *and an integer* k_0 *depending on it such that for* $\lambda \geq \lambda_0$ *the system* (5.1), (5.2) *has strongly unstable* k*-cycles with indices* $0 \leq k \leq k_0$.

These assertions, which follow immediately from our constructions, enable us to explain in full scope the results of numerical experiments with the system (5.1), (5.2). This has to do with the fact that for the values $\lambda \leq 6$ for which numerical integration of the system (5.1), (5.2) was carried out on the whole, and for relatively small a the parameter a_0 in (5.88) is also not large.

Putting everything together, we conclude that for finite a our asymptotic theory has a limiting character: it works for values of the Malthusian coefficient of linear growth of the prey that are so large that the biological content of the problem is essentially lost.

In concluding our treatment of the dynamics of the system (5.1), (5.2) when $\Delta > 1$, we present intuitive considerations allowing us to conclude that in Theorems 5.1–5.4 attractors with a broad domain of attraction are constructed. Suppose that the initial conditions of the system are given for $t \leq t_{\text{in}}$. Assume that for these t the initial function $N_1(t)$ has the form (5.3) with $b = 1$, while the second component of the initial condition is a monotonically increasing continuous function $N_2(t)$, also depending on the large parameter λ, such that

$$\int_{t_{\text{in}}-h}^{t_{\text{in}}} N_2(s)\, ds = o(1), \qquad N_2(t_{\text{in}}) = 1.$$

Obviously, t_{in} plays the role of t_0. Therefore, for $t \geq t_{\text{in}}$ equation (5.29) can be used to construct $N_1(t)$ and $N_2(t)$, and hence the approximation formula (5.37) is valid, with

$$N_2(t_1) = \frac{r(1+a)}{a\lambda} N_1(t_{\text{in}}). \tag{5.89}$$

We proceed to the next spike in the prey density. Changing the time reference point, we can assume that the time begins at $t = -1$. We conclude from (5.3) with $b = 1$ that, according to (5.37), the approximate equality

$$N_2(t) = N_2(t_1)\exp\left[-rhN_2(t_1) + r(1+a)\lambda^{-1}\exp\lambda\left(1 - \exp\left(-\frac{\exp\lambda t}{1+a}\right)\right)\right] \tag{5.90}$$

holds for $-1 \le t \le t_{\text{out}}$ and $N_2(t_{\text{out}}) = 1$. It follows without difficulty from (5.89), (5.90), and (5.3) with $b = 1$ that, approximately,

$$t_{\text{out}} = \frac{1}{\lambda}\ln\left[-(1+a)\ln\left(1 - \frac{rh}{a}\exp\left(\lambda t_{\text{in}} - \frac{\exp\lambda t_{\text{in}}}{1+a}\right)\right)\right]. \tag{5.91}$$

Suppose that $t_{\text{in}} = \varphi/\lambda$ in (5.91). Using this and making the substitution

$$\varphi = \ln[-(1+a)\ln z],$$

we arrive at the mapping (5.41). Of basic interest to us is the case when in (5.91)

$$t_{\text{in}} = -\alpha + \beta/\lambda, \tag{5.92}$$

where $0 < \alpha < 1$, and the sign of the parameter β is arbitrary. Under the condition (5.92) we conclude from (5.91) that, approximately,

$$\beta \to \beta + \ln\Delta. \tag{5.93}$$

It follows from (5.93) that for $\Delta > 1$ the iterations of the mapping (5.93) go to infinity. By using (5.91) it can be shown that under the condition (5.92) we get

$$\beta = \alpha\lambda + O(1) \tag{5.94}$$

in λ steps. We conclude from a comparison of (5.92) and (5.94) that t_{in} becomes of order λ^{-1}, where the mapping (5.41) works.

Thus, the mapping (5.91) and the mapping (5.41) following from it are obtained under sufficiently general assumptions about the character of behavior of solutions of the system (5.1), (5.2), in particular, actually without the use of the conceptually laden Lemmas 5.1–5.4. In other words, we want to say that the mapping (5.41) is obtained if we use only the most general properties of solutions of the system (5.1), (5.2). This remark is made in connection with [20], where the mapping (5.41) was derived much less rigorously than above, but which does not contain a solution of the problem considered by us, of course.

If $\Delta < 1$, then there is a departure from oscillations of Hutchinson type. In the next subsection (written jointly by the authors) we reveal the character of steady-state regimes of the system (5.1), (5.2) under this restriction in an important special case.

5.5. On the dynamics of the system (5.1), (5.2) for small values of the coefficient of linear growth of the predator. To give an idea of the dynamics of the system (5.1), (5.2) for $\Delta < 1$ we study the case when

$$r = \lambda\exp(-\alpha\lambda), \qquad \alpha > 0, \tag{5.95}$$

which is interesting from many points of view. To simplify the exposition we omit the proofs of the C^1-assertions, confining ourselves to an outline of the investigation.

We begin our analysis with asymptotic integration of the system (5.1), (5.2) for the initial functions

$$N_1(t) = \exp\lambda(1 - ax_0/(1+a))t\,, \qquad N_2(t) \equiv x_0\,, \tag{5.96}$$

which are defined for $t \le 0$. In (5.96) x_0 is a constant in the interval $(0, 1 + 1/a)$. Suppose first that

$$x_0 < (1 + 1/a)(1 - \alpha)\,. \tag{5.97}$$

Since r is small, the variables in the system (5.1), (5.2) interact weakly under the condition (5.97) and for $0 \le t \le t_0 - \delta$, where $t_0 = \alpha/(1 - ax_0/(1+a))$ and the constant $\delta > 0$ is sufficiently small. This implies that, uniformly with respect to the indicated t,

$$N_1(t) = [1 + o(1)]\exp\lambda(1 - ax_0/(1+a))t\,, \tag{5.98}$$

$$N_2(t) = [1 + o(1)]x_0 \exp\left[\frac{1+a}{1+(1-x_0)a}\exp\lambda\left(1 - \frac{ax_0}{1+a}\right)(t - t_0)\right]. \tag{5.99}$$

But if $|t - t_0| \le \delta$, i.e., for strong interaction of the variables, the leading asymptotics of N_1 and N_2 are found by solving the system

$$\dot N_1 = \lambda(1 - a(1+a)^{-1}N_2)N_1\,, \qquad \dot N_2 = rN_1N_2\,. \tag{5.100}$$

The proof is as follows: since $N_2(t)$ is bounded, the term $rN_2(t-h)N_2(t)$ in (5.2) can be omitted, and since $t_0 < 1$ by (5.97), the term containing the delay in (5.1) can also be discarded. In (5.100) we set

$$y = r\int_0^t N_1(s)\,ds\,, \qquad x_0\exp y = N_2\,, \qquad \tau = \lambda(t - t_0)\,.$$

As a result, we get the equation (cf. (5.27))

$$\ddot y = [1 - ax_0(1+a)^{-1}\exp y]\dot y\,, \tag{5.101}$$

which should be considered for $-\infty < \tau < \infty$. According to (5.98), we are interested in a solution of it such that as $\tau \to -\infty$

$$y \sim \frac{1}{1 - ax_0/(1+a)}\exp\left(1 - \frac{ax_0}{1+a}\right)\tau\,, \qquad \dot y \sim \exp\left(1 - \frac{ax_0}{1+a}\right)\tau\,. \tag{5.102}$$

Using (5.102) and integrating the left-hand and right-hand sides of (5.101) over $(-\infty, \tau]$, we arrive at the equation

$$\dot y = y - ax_0(1+a)^{-1}(\exp y - 1)\,, \tag{5.103}$$

from which it is convenient to pass to the integral equality

$$\begin{aligned} y(\tau) = {}& \frac{1+a}{1+(1-x_0)a} \\ & \times \exp\left[\frac{1+(1-x_0)a}{1+a}\tau + \frac{ax_0}{1+a}\int_{-\infty}^{\tau}(1 + y(s) - \exp y(s))\,ds\right]. \end{aligned} \tag{5.104}$$

In the Banach space $V(\tau_0)$ consisting of continuous functions on $(-\infty, -\tau_0]$, with the norm

$$\|y\|_{\tau_0} = \sup_{\tau \le -\tau_0}\left|y(\tau)\exp\left[-\frac{1+(1-x_0)a}{1+a}\tau\right]\right|,$$

we single out the set

$$S = \left\{ y(\tau) \in V(\tau_0) : y(\tau) \geq 0 , \|y\|_{\tau_0} \leq \frac{1+a}{1+(1-x_0)a} \right\} . \tag{5.105}$$

It is not hard to show that the operator generated by the right-hand side of (5.104) transforms the set (5.105) into itself and is contracting on this set when τ_0 is a sufficiently large positive number. We extend the function $y(\tau)$ to all values of τ according to (5.103), which gives us that the function is monotonically increasing and tends to a finite limit $y_0(x_0)$ as $\tau \to \infty$, where y_0 is the positive root of the equation

$$ax_0(1+a)^{-1}(\exp y - 1) = y . \tag{5.106}$$

It follows from our constructions that the function rN_1 has an exponentially small value at $t = t_0 + \delta$, it keeps this property up to the time $t = 1$, and it falls to superexponentially small values for $t > 1$. We note also that the value of N_2 at $t = t_0 + \delta$ is asymptotically close to $x_0 \exp y_0$.

Denote by t_1 the second positive root of the equation $N_1(t) = 1$. To find it asymptotically we observe that after the substitution $rt = s$ equation (5.2) can be approximated with exponential accuracy by the equation $\dot{N}_2 = -N_2^2$ on any finite interval of variation of s. Supplementing this equation by the initial condition $N_2(0) = x_0 \exp y_0$, we find that

$$N_2(s) = x_0 \exp y_0/(1 + sx_0 \exp y_0) . \tag{5.107}$$

If we then use (5.107) in the equality

$$t_1 - \frac{a}{1+a} \int_0^{t_1} N_2(t)\, dt - \frac{1}{1+a} \int_{-1}^{t_1-1} N_1(t)\, dt = 0 ,$$

we see that

$$t_1 = [1 + o(1)] \frac{s_0}{\lambda} \exp \alpha\lambda , \tag{5.108}$$

where s_0 is the root of the equation

$$s_0 - \frac{a}{1+a} \ln(1 + s_0 x_0 \exp y_0) = \frac{y_0}{1+a} . \tag{5.109}$$

We introduce the mapping

$$x_0 \to f(x_0) = x_0 \exp y_0/(1 + s_0 x_0 \exp y_0) . \tag{5.110}$$

It follows from the asymptotic properties of N_1 and N_2 noted above that the following relations hold uniformly with respect to $-1 \leq t \leq 0$ and $-h \leq t \leq 0$ (cf. (5.96)):

$$N_1(t + t_1) = [1 + o(1)] \exp \lambda \left(1 - \frac{af(x_0)}{1+a}\right) t , \tag{5.111}$$

$$N_2(t + t_1) = [1 + o(1)] f(x_0) . \tag{5.112}$$

Therefore, it makes sense to dwell separately on the properties of the mapping (5.110).

LEMMA 5.5. *The mapping* (5.110) *transforms the interval* $(0, 1+1/a)$ *into itself, is monotone on that interval, and has a unique fixed point* $x_* < 1$, *with* $0 < f'(x_*) < 1$.

PROOF. We first establish the monotonicity of the mapping (5.110). With this purpose we introduce the function $x_1(x_0) = x_0 \exp y_0(x_0)$. From (5.106) and the representation $y_0 = \ln x_1 - \ln x_0$ we derive successively

$$x_1'(x_0) = x_1\left(1 - \frac{ax_0}{1+a}\right) \Big/ x_0\left(1 - \frac{ax_1}{1+a}\right) < 0, \tag{5.113}$$

$$y_0'(x_0) = a(x_1 - x_0)/x_0(1+a)\left(1 - \frac{ax_1}{1+a}\right) < 0. \tag{5.114}$$

To clarify the situation with the inequalities (5.113) and (5.114) we note that

$$ax_1/(1+a) > 1. \tag{5.115}$$

But the inequality (5.115) is obvious, since it means that $\varphi'(y_0) > 0$, where $\varphi(y) = ax_0(1+a)^{-1}(\exp y - 1) - y$, and $\varphi(y_0) = 0$.

To advance further we consider the function $x_2 = x_2(x_1)$ determined from the equation

$$-\int_{x_1}^{x_2}\left(1 - \frac{a}{1+a}s\right)\frac{ds}{s^2} = \frac{y_0(x_0)}{1+a}, \tag{5.116}$$

in which $x_0 = x_0(x_1)$ (this is all right because of (5.113)). From (5.116), (5.113)–(5.115), and the estimate $ax_2/(1+a) < 1$ (it can be obtained from (5.116) similarly to the derivation of (5.115) it follows that

$$x_2'(x_1) = \frac{x_2^2}{ax_2/(1+a) - 1}\left\{\frac{1}{x_1^2}\left(\frac{ax_1}{1+a} - 1\right) + \frac{y_0'(x_0)x_0'(x_1)}{1+a}\right\} < 0. \tag{5.117}$$

Finally, we conclude from (5.113), (5.117), and the equality $f(x_0) = x_2$ that the mapping (5.110) transforms the interval $(0, 1 + 1/a)$ into itself, with

$$f'(x_0) = x_2'(x_1)x_1'(x_0) > 0, \qquad 0 < x_0 < 1 + 1/a. \tag{5.118}$$

To find the fixed point of the mapping (5.110) we use the representation

$$f(x_0) = x_0 \exp a^{-1}(1+a)(y_0 - s_0),$$

which follows from (5.109) and which implies that $y_0 = s_0$ on the fixed point. Therefore, if in (5.109) we set $s_0 = y_0$ and (see (5.106))

$$x_0 = a^{-1}(1+a)y_0/(\exp y_0 - 1), \tag{5.119}$$

we arrive at the equation

$$y_0 = 2\sqrt{a(1+a)^{-1}}\sinh(y_0/2) \tag{5.120}$$

for finding y_0, which obviously has a unique positive root $y_0 = y_*$. Thus, the mapping (5.110) has a unique fixed point on $(0, 1 + 1/a)$, and for it we get the following formula by setting $s_0 = y_*$ in (5.109):

$$x_* = [\exp y_* - 1]/y_* \exp y_* < 1.$$

By (5.118) it remains to show that $f'(x_*) < 1$. We note that the uniqueness of the fixed point implies only that $f'(x_*) \le 1$. However, if $f'(x_*) = 1$, then $s_0'(x_*) =$

$y_0'(x_*)$. And from this we conclude after some transformations that $y_* = y_0(x_*)$ satisfies the equality

$$\frac{a}{(1+a)^2}\frac{\exp y_* - 1}{1 - \exp y_* + y_* \exp y_*} + \frac{(\exp y_* - 1 - y_*)^2}{(\exp y_* - 1)^2} = 0. \tag{5.121}$$

The latter is not possible, since the left-hand side of (5.121) is positive for all $y_* > 0$. The lemma is proved.

We note in addition that $f(x_0) \to 0$ or $1 + 1/a$ as $x_0 \to +0$ or $x_0 \to 1 + 1/a - 0$, respectively, and $f'(x_0) \to \infty$ in both cases.

THEOREM 5.7. *Suppose α is an arbitrarily fixed constant in the interval $(0, \alpha_*)$, where $\alpha_* = 1 - ax_*(1+a)^{-1}$, and x_* is the fixed point of the mapping* (5.110). *Then there is a sufficiently large $\lambda_0 > 0$ such that for all $\lambda \geq \lambda_0$ the system* (5.1), (5.2) *has under the condition* (5.95) *an exponentially orbitally stable cycle whose initial conditions and whose period satisfy the asymptotic formulas* (5.111), (5.112), *and* (5.108) (*with $f(x_0)$ replaced by x_*, and s_0 by $s_* = s_0(x_*)$*).

The proof of the existence of the cycle is based on properties of the solutions of the system (5.1), (5.2) with initial conditions (5.96) and on Lemma 5.5 (see, for example, Theorems 5.1 and 5.2), with the same constructions as in nonlinear analysis (see §5.3) used in the asymptotic integration of the variational system. In particular, a linear nonhomogeneous integral equation analogous to (5.104) arises on the strong interaction part. Omitting the computations, we note only that as $\lambda \to \infty$ all the multipliers of the variational system tend to zero except for two—the simple unit and one other having $f'(x_*)$ as its limit.

The behavior of the solutions of the system (5.1), (5.2) with initial conditions (5.96) for which (cf. (5.97))

$$(1-\alpha)(1+1/a) < x_0 < 1 + 1/a \tag{5.122}$$

reveals itself in a simple way, since in this case the variables in (5.1), (5.2) interact weakly: $N_2(t)$ is asymptotically close to x_0 for all the t under consideration, and $N_1(t)$ varies according to the Hutchinson law. Therefore, here (cf. (5.108))

$$t_1 = [1 + o(1)]\frac{1+a}{\lambda(1 + (1-x_0)a)} \exp \lambda \left(1 - \frac{ax_0}{1+a}\right), \tag{5.123}$$

and instead of (5.111) and (5.112) we have the following, uniformly with respect to $-1 \leq t \leq 0$ and $-h \leq t \leq 0$:

$$N_1(t + t_1) = [1 + o(1)] \exp \lambda \left(1 - \frac{ax_0}{1+a}\right) t, \tag{5.124}$$

$$N_2(t+t_1) = x_0 + \left[\frac{(1+a)^2(1-x_0)x_0}{1+(1-x_0)a} + o(1)\right] \exp\left(-\lambda\left(a + \frac{ax_0}{1+a} - 1\right)\right). \tag{5.125}$$

It is now clear enough that under the condition (5.122) the system (5.1), (5.2) has a cycle such that the constant x_0 is asymptotically close to 1. We formulate the precise statement.

THEOREM 5.8. *Let α be an arbitrarily fixed constant in the interval* $(1/(1+a), \infty)$. *Then for all sufficiently large λ the system* (5.1), (5.2) *has under the condition* (5.95) *an exponentially orbitally stable cycle whose period and initial conditions satisfy the asymptotic formulas* (5.123) *and* (5.124), (5.125) *with* $x_0 = 1$.

Without dwelling on the standard proof of the theorem, we mention only that all the multipliers of the variational system on the cycle tend to zero with the exception of the simple unit and a multiplier μ satisfying the asymptotic formula

$$\mu = 1 - [1 + o(1)](1 + a)^2 \exp[-\lambda(\alpha - 1/(1 + a))].$$

We mention also that with increasing r the cycle in Theorem 5.8 branches off from the exponentially orbitally stable continuum of periodic solutions that exists for $r = 0$.

As will be clear from what follows, we achieve a certain completeness of the investigation if we study the case when the x_0 in (5.96) is replaced by

$$x = x_0 + (1 + a)\gamma/a\lambda, \qquad -\infty < \gamma < \infty, \tag{5.126}$$

and $\alpha = 1 - ax_0/(1 + a)$ in (5.95). Integrating the system (5.1), (5.2) with initial conditions modified as indicated, we get the following formulas (cf. (5.98), (5.99)) for $0 \le t \le 1 - \delta$, where $\delta > 0$ is sufficiently small and fixed:

$$N_1(t) = [1 + o(1)] \exp \lambda \left(1 - \frac{ax}{1 + a}\right) t, \tag{5.127}$$

$$N_2(t) = [1 + o(1)]x_0 \exp \left\{\frac{1 + a}{1 + (1 - x_0)a} \exp \left[\lambda \left(1 - \frac{ax}{1 + a}\right)(t - 1) - \gamma\right]\right\}. \tag{5.128}$$

On the strong interaction part (that is, for $|t - 1| \le \delta$), we now have in place of (5.100) the system

$$\dot{N}_1 = \lambda \left[1 - \frac{a}{1 + a} N_2 - \frac{1}{1 + a} \exp \lambda \left(1 - \frac{ax}{1 + a}\right)(t - 1)\right] N_1, \tag{5.129}$$

$$\dot{N}_2 = rN_1N_2. \tag{5.130}$$

Making in (5.129) and (5.130) the substitutions

$$y = r\int_0^t N_1(s)\,ds, \qquad x \exp y = N_2, \qquad \tau = \lambda(t - 1),$$

and then setting $x = x_0$, we obtain

$$\ddot{y} = \left[1 - \frac{ax_0}{1 + a} \exp y - \frac{1}{1 + a} \exp \left(1 - \frac{ax_0}{1 + a}\right)\tau\right] \dot{y}, \tag{5.131}$$

for which the solution of interest to us must have the following properties by virtue of (5.127): as $\tau \to \infty$

$$y \sim \frac{c(1 + a)}{1 + (1 - x_0)a} \exp \left(1 - \frac{ax_0}{1 + a}\right)\tau, \qquad \dot{y} \sim c \cdot \exp \left(1 - \frac{ax_0}{1 + a}\right)\tau, \tag{5.132}$$

where $c = \exp(-\gamma)$.

We set $z(\tau) = \dot{y}(\tau)$ and pass from (5.131) with the use of (5.132) to the integral equation

$$z(\tau) = c \cdot \exp\left\{\left(1 - \frac{ax_0}{1+a}\right)\tau + \frac{ax_0}{1+a}\int_{-\infty}^{\tau}\left[1 - \exp\left(\int_{-\infty}^{s} z(\sigma)\,d\sigma\right)\right] ds \right. \\ \left. - \frac{1}{1+(1-x_0)a}\exp\left(1 - \frac{ax_0}{1+a}\right)\tau\right\}, \tag{5.133}$$

for which the analysis is completely analogous to the investigation of (5.104). In this way we get a function $z(\tau)$ defined for all $-\infty < \tau < \infty$ and tending to zero with superexponential accuracy as $\tau \to \infty$. Returning again to (5.131), we see that its desired solution

$$y(\tau) = \int_{-\infty}^{\tau} z(s)\,ds$$

is defined for all $-\infty < \tau < \infty$, is monotonically increasing, and tends as $\tau \to \infty$ to the finite limit

$$z_0(x_0, \gamma) = \int_{-\infty}^{\infty} z(\tau)\,d\tau.$$

This analysis shows that for $t = 1+\delta$ the function N_1 becomes superexponentially small, while the value of N_2 is asymptotically close to $x_0 \exp z_0$. Therefore, the subsequent investigation coincides completely with the exposition at the beginning of the subsection: for the second positive root of the equation $N_1(t) = 1$ we have the asymptotic representation

$$t_1 = [1 + o(1)]\frac{\sigma_0}{\lambda}\exp\lambda\left(1 - \frac{ax_0}{1+a}\right), \tag{5.134}$$

where σ_0 is the root of the equation

$$(1+a)\sigma_0 - a\ln(1 + \sigma_0 x_0 \exp z_0) = z_0, \tag{5.135}$$

and instead of (5.111) and (5.112) we get the following, uniformly with respect to $-1 \le t \le 0$ and $-h \le t \le 0$:

$$N_1(t + t_1) = [1 + o(1)]\exp\lambda(1 - a\tilde{x}(1+a)^{-1})t, \tag{5.136}$$

$$N_2(t + t_1) = [1 + o(1)]\tilde{x}, \tag{5.137}$$

where

$$\tilde{x} = x_0 \exp z_0/(1 + \sigma_0 x_0 \exp z_0). \tag{5.138}$$

Therefore, it is natural to expect that under the condition (5.126) a cycle of the original system (5.1), (5.2) must correspond to each simple root γ_0 of the equation

$$G(x_0, \gamma) = x_0 \tag{5.139}$$

for a fixed $x_0 \in (0, 1 + 1/a)$, where G is the right-hand side of (5.138). It is also clear that the mapping

$$\gamma \to \lambda a(1+a)^{-1}[G(x_0, \gamma) - x_0] \tag{5.140}$$

must be responsible for the stability of the cycle. And since this mapping has an exponentially unstable fixed point $\gamma(\lambda) = \gamma_0 + O(\lambda^{-1})$, the relaxation cycle of the system (5.1), (5.2) will be the same.

To make the subsequent constructions precise, we dwell in greater detail on the properties of the function $G(x_0, \gamma)$. It follows from (5.133) that the asymptotic representation

$$z_0 = (1+a)c + O(c^2) \tag{5.141}$$

is valid as $\gamma \to \infty$, i.e., as $c \to 0$. Using (5.141) in (5.135), we conclude that

$$\sigma_0 = \frac{c}{1 - ax_0/(1+a)} + O(c^2). \tag{5.142}$$

Finally, substituting (5.141) and (5.142) into (5.138), we get that for large γ

$$G(x_0, \gamma) = x_0 \left[1 + \frac{(1+a)(1-x_0)}{1 - ax_0/(1+a)} \exp(-\gamma) + O(\exp(-2\gamma))\right]. \tag{5.143}$$

Making the substitution $\tau = \sigma + \gamma/(1 - ax_0/(1+a))$ in (5.133) and letting γ go to $-\infty$, we see that $z_0 \to y_0(x_0)$. Therefore,

$$G(x_0, \gamma) \to f(x_0), \qquad \gamma \to -\infty, \tag{5.144}$$

where f is the function (5.110).

LEMMA 5.6. *The respective inequalities* $G(x_0, \gamma) > x_0$ *and* $G(x_0, \gamma) < x_0$, $-\infty < \gamma < \infty$, *hold for* $0 < x_0 < x_*$ *and* $1 < x_0 < 1 + 1/a$, *where* x_* *is the fixed point of the mapping* (5.110). *But if* $x_* < x_0 < 1$, *then equation* (5.139) *has a unique root* $\gamma_0 = \gamma_0(x_0)$, *and*

$$\frac{\partial G}{\partial \gamma}(x_0, \gamma_0) > 0. \tag{5.145}$$

PROOF. By (5.143) and (5.144) and the properties of f established earlier, to prove the lemma it suffices to prove that if equation (5.139) has the root γ_0, then the inequality (5.145) necessarily holds.

Let γ_0 be a root of equation (5.139). It follows from the representation $G = x_0 \exp a^{-1}(1+a)(z_0 - \sigma_0)$ that $\sigma_0 = z_0$ for $\gamma = \gamma_0$. Using this equality in (5.135), we find that

$$x_0 = (\exp z_0 - 1)/z_0 \exp z_0, \tag{5.146}$$

$$\frac{\partial \sigma_0}{\partial \gamma} = \left(\frac{a}{1+a} x_0 z_0 + \frac{1}{1+a}\right) \frac{\partial z_0}{\partial \gamma} \Big/ \left(1 - \frac{ax_0}{1+a}\right). \tag{5.147}$$

If we then substitute (5.146) and (5.147) in the formula

$$\frac{\partial G}{\partial \gamma} = x_0 \frac{1+a}{a} \left(\frac{\partial z_0}{\partial \gamma} - \frac{\partial \sigma_0}{\partial \gamma}\right),$$

we get finally that

$$\frac{\partial G}{\partial \gamma}(x_0, \gamma_0) = -\frac{x_0}{1 - ax_0/(1+a)} \frac{\exp z_0 - 1 - z_0}{z_0 \exp z_0} \frac{\partial z_0}{\partial \gamma}(x_0, \gamma_0).$$

Thus, the problem has been reduced to the proof of the inequality $\partial z_0/\partial \gamma < 0$, or, what is the same, $\partial z_0/\partial c > 0$.

Let $h_0(\tau) = \frac{\partial z}{\partial c}(\tau)$. Recall that

$$\frac{\partial z_0}{\partial c} = \int_{-\infty}^{\infty} h_0(\tau)\, d\tau.$$

Therefore, it suffices to show that the function $h_0(\tau)$ is positive.

It follows from (5.133) that h_0 is a solution of the linear nonhomogeneous equation

$$h_0(\tau) + \frac{ax_0}{1+a} z(\tau) \int_{-\infty}^{\tau} \omega(s) \int_{-\infty}^{s} h_0(\sigma)\, d\sigma\, ds = \frac{1}{c} z(\tau), \tag{5.148}$$

where

$$\omega(\tau) = \exp\left(\int_{-\infty}^{\tau} z(s)\, ds\right).$$

We introduce the operator

$$Lh_0 = \frac{ax_0}{1+a} z(\tau) \int_{-\infty}^{\tau} \omega(s) \int_{-\infty}^{s} h_0(\sigma)\, d\sigma\, ds.$$

Recall that the function $z(\tau)$ is positive, and as $\tau \to -\infty$ and ∞ it tends to zero at exponential and superexponential rates, respectively. From this and from the structure of the operator L we see that it acts in the space $V(\tau_0)$ and is compact for any $\tau_0 \in (-\infty, \infty)$. But since it is a Volterra operator, its spectral radius is equal to zero. Therefore, equation (5.148) is uniquely solvable. To prove that $h_0(\tau)$ is positive, we divide equation (5.148) by $z(\tau)$ and write the result in the form

$$u_0(\tau) + \frac{ax_0}{1+a} \int_{-\infty}^{\tau} \omega(s) \int_{-\infty}^{s} z(\sigma) u_0(\sigma)\, d\sigma\, ds = \frac{1}{c}, \tag{5.149}$$

where $u_0(\tau) = h_0(\tau)/z(\tau)$. It follows from (5.149) that there is a sufficiently large positive number τ_0 such that $u_0(\tau) > 0$, $-\infty < \tau \le -\tau_0$. It remains to use once more the Volterra property of the integral operator in (5.149) in order to see by a contradiction argument that $u_0(\tau)$ is positive for all τ. We arrive at the required result: the integral of $h_0(\tau)$ over $-\infty < \tau < \infty$ is finite and positive. The lemma is proved.

We now formulate the rigorous result.

THEOREM 5.9. *Suppose that for a fixed* $x_0 \in (x_*, 1)$

$$\left|\frac{\partial G}{\partial x_0}(x_0, \gamma_0)\right| \neq 1, \tag{5.150}$$

where γ_0 *is the root of equation* (5.139). *Then for all sufficiently large* λ *there is a unique function* $\gamma = \gamma(\lambda)$ *with* $\gamma(\infty) = \gamma_0$ *such that the system* (5.1), (5.2) *with*

$$r = \lambda \exp\left(-\lambda\left(1 - \frac{ax_0}{1+a}\right) - \gamma(\lambda)\right)$$

has a strongly unstable cycle, and the following asymptotic formulas hold for the initial conditions, the period $T(\gamma)$, *and the leading (greatest in modulus) multiplier:*

$$N_1(t, \lambda) = [1 + o(1)] \exp \lambda \left(1 - \frac{ax_0}{1+a}\right) t, \qquad N_1(0, \lambda) = 1, \qquad -1 \le t \le 0,$$

$$N_2(t, \lambda) = [1 + o(1)] x_0, \qquad N_2(0, \lambda) = x_0, \qquad -h \le t \le 0,$$

$$T(\lambda) = [1 + o(1)] \frac{\sigma_0}{\lambda} \exp \lambda \left(1 - \frac{ax_0}{1+a}\right), \qquad \sigma_0 = \sigma_0(x_0, \gamma_0),$$

$$\mu = [1 + o(1)] \frac{\lambda a}{1+a} \frac{\partial G}{\partial \gamma}(x_0, \gamma_0).$$

To prove that there is a cycle we use the following special device. For the system (5.1), (5.2) we consider the set $\Omega_1 \times \Omega_2$ of initial functions, where

$$\Omega_1 = \{N_1(t) \in C(-1, 0) : 0 \leq N_1(t) \leq \exp \lambda m_0 t, N_1(0) = 1\},$$
$$\Omega_2 = \{N_2(t) \in C(-h, 0) : x_0 - \delta \leq N_2(t) \leq x_0 + \delta\},$$

and $m_0, \delta > 0$ are sufficiently small universal constants, and in (5.2) we set

$$r = \lambda \exp(-\lambda(1 - ax(1+a)^{-1}) - \gamma), \tag{5.151}$$

where the parameter γ varies in a small neighborhood of γ_0, and $x = N_2(0)$. The equality (5.151) means that r is now a functional of the initial condition of $N_2(t)$. Integrating the system (5.1), (5.2) asymptotically with the indicated initial conditions (see the corresponding place before Lemma 5.6 for the technical details), we arrive at (5.136) and (5.137), in which now $\tilde{x} = G(x, \gamma)$.

Thus, the problem of the existence of a cycle reduces to the determination of a structurally stable fixed point of the mapping $x \to G(x, \gamma)$. It remains to note that, by the choice of γ_0 and the condition (5.150), such a fixed point $x = x(\gamma)$, $x(\gamma_0) = x_0$, is unique in a neighborhood of x_0. This proves the existence of a cycle $N_1(t, \lambda, \gamma)$, $N_2(t, \lambda, \gamma)$ for the system (5.1), (5.2) with

$$r = \lambda \exp\{-\lambda[1 - a(1+a)^{-1} N_2(0, \lambda, \gamma)] - \gamma\}$$

for all γ close to γ_0. The concluding stage is the determination of $\gamma = \gamma(\lambda)$ from the condition $N_2(0, \lambda, \gamma) = x_0$, which is possible by virtue of Lemma 5.6, the asymptotic equalities

$$N_2(0, \lambda, \gamma) = x(\gamma) + o(1), \qquad \frac{\partial N_2}{\partial \gamma}(0, \lambda, \gamma) = x'(\gamma) + o(1),$$

and the inequalities (see (5.145) and (5.150))

$$x'(\gamma_0) = \frac{\partial G}{\partial \gamma}(x_0, \gamma_0) \Big/ \left(1 - \frac{\partial G}{\partial x_0}(x_0, \gamma_0)\right) \neq 0.$$

We do not dwell on the proof that the cycle is unstable. We remark only that the remaining multipliers (besides 1 and μ) of the variational system tend to zero as $\lambda \to \infty$.

We point out that an analogous device, the so-called method of parameter functionalization, was used previously in [24] for constructing unstable duck cycles in multidimensional relaxation systems of ordinary equations, and a restriction analogous to (1.151) also arose there.

Theorems 5.7–5.9 allow us to follow the dynamics of the system (5.1), (5.2) under the condition (5.95) as α decreases. Indeed, for all $\alpha > \alpha_*$ the system (5.1), (5.2) has an orbitally exponentially stable cycle supplied by Theorem 5.8 which appears to be a global attractor of it. As α passes through α_* two more cycles are born from condensation of trajectories: a stable cycle (supplied by Theorem 5.7) and an unstable cycle (supplied by Theorem 5.9). For $(1+a)^{-1} < \alpha < \alpha_*$ all three cycles coexist, and as α passes through $(1+a)^{-1}$ the cycles supplied by Theorems 5.8 and 5.9 consolidate and disappear. Thus, for $0 < \alpha < (1+a)^{-1}$ only the stable cycle in Theorem 5.7 survives.

The last cycle is probably preserved for all r such that $\Delta < 1$. In particular, it is likely that the cycle in Theorem 5.2 is transformed into it as Δ passes through 1.

5.6. Summary remarks. Thus, we have constructed a rather elaborate asymptotic theory of periodic relaxation oscillations of the system (5.1), (5.2). It is essential that for $\Delta > 1$ a stochastization of them [69] can take place, which, however, bears a sufficiently regular character. It is a nontrivial circumstance that our theory has a limit character (this follows most clearly from Theorems 5.5 and 5.6). This statement is probably typical, though, for systems of differential equations modeling turbulence phenomena. As an example we mention the well-known system of Lorenz, for which only cycles with relatively simple structure remain from compound oscillations when the energy parameter is increased.

6. The relaxation cycle of a mathematical model of the Belousov reaction

6.1. The mathematical model and some of its properties. Belousov discovered the reaction of interest to us in 1951, but his detailed paper on this topic was not published until fairly recently [2]. There is a thorough experimental analysis of the Belousov reaction in [18]. At present many oxidation-reduction chemical reactions are known with the same type of mechanism. Following [18], we describe this mechanism, in which bromate, cerium, and bromide are singled out as the basic substances. On a qualitative level the mechanism of the oscillations is as follows: the cerium ($[Ce^{4+}] = N_2$), which plays the role of a catalyst, promotes the formation of bromide ($[Br^-] = N_3$), which serves as a strong inhibitor of bromate ($[HBrO_2] = N_1$), while the latter increases the concentration of cerium.

In [18] the period of auto-oscillations of cerium is divided into time intervals τ_1 and τ_2, the phases of increase and fall of the concentration, respectively, which permits us to formulate the following statements.

(a) Small amounts of Br^-, added in the phase τ_1, cause a switch, that is, a transition to the decrease phase τ_2. The minimal amount of Br^- causing a switch decreases by the end of the phase τ_1.

(b) Br^- added in the phase τ_2 lengthens it. The effect is all the stronger, the greater the addition and the closer it is to the end of the phase τ_2.

(c) An addition of Ag^+, which binds the Br^-, causes the following effects: if the concentration of Br^- is decreased in the phase τ_2, then a switch ensues, and an addition of Ag^+ in the phase τ_1 leads to an increase in its duration.

(d) The auto-oscillations disappear if the concentration of Br^- is increased continuously.

(e) A result analogous to the preceding is obtained if the concentration of Ce^{4+} is increased.

(f) An addition of Ce^{4+} in the phase τ_1 leads to a switch.

Comparing these facts with the general approach to modeling of biological populations in §1.2, we conclude that the Belousov reaction is surprisingly similar to a predator-prey problem of some kind. In this connection we introduce the following system of ordinary differential equations to model it:

$$\dot{N}_1 = r_1[1 + a(1 - N_3/K_3) - N_1/K_1]N_1, \tag{6.1}$$

$$\dot{N}_2 = r_2[N_1/K_1 - N_2/K_2]N_2, \tag{6.2}$$

$$\dot{N}_3 = r_3[N_2/K_2 - N_3/K_3]N_3, \tag{6.3}$$

where K_s and r_s $(s = 1, 2, 3)$ are the average concentrations and the coefficients of linear growth of the bromate, cerium, and bromide, respectively. In §1.2 we mentioned that some of the parameters in the ecological equations are invariants of them. In this case it can be expected that the quantities

$$r_2/K_2 \quad \text{and} \quad r_3/K_3 \tag{6.4}$$

are preserved when nonspecific conditions of the course of the reaction change. The intuitive meaning here is as follows: if the average concentrations of cerium or bromide are increased, then on the average the rates of change of their concentrations increase in a corresponding way. We assume below that this fairly reasonable assumption holds.

We linearize the system (6.1)–(6.3) on its equilibrium state

$$(K_1, K_2, K_3). \tag{6.5}$$

Then we form the characteristic equation

$$\mu/r_1 = -1 - ar_2r_3/[(\mu + r_2)(\mu + r_3)]. \tag{6.6}$$

From (6.6) we easily derive the following assertion.

THEOREM 6.1. *Suppose that the biquadratic equation*

$$\omega^4 - (ar_2r_3 - r_2^2 - r_3^2)\omega^2 + (1 + a)r_2^2r_3^2 = 0 \tag{6.7}$$

has roots $\omega_2 > \omega_1 > 0$. *Then the equilibrium state* (6.5) *of the system* (6.1)–(6.3) *is unstable in an oscillatory way for*

$$\frac{(\omega_2^2 + r_2^2)(\omega_2^2 + r_3^2)}{ar_2r_3(r_2 + r_3)} > r_1 > \frac{(\omega_1^2 + r_2^2)(\omega_1^2 + r_3^2)}{ar_2r_3(r_2 + r_3)}. \tag{6.8}$$

Figure 6.1 shows the evolution of the oscillations in the cerium density for $a = 40$, $r_2 = 1$, and $r_3 = 3$ when

$$r_1 = 0.1; 0.15; 0.2; 0.3; 1; 10.$$

We note also that in the overwhelming majority of cases the auto-oscillations in the system (6.1)–(6.3) arise strictly.

We pass to an explanation of the experimental facts given above.

(a) Suppose that N_3 is increased in the phase τ_1. Then N_1 decreases by (6.1), which leads to a decrease in N_2 according to (6.2). The duration of the phase τ_1 is thereby shortened. Obviously, the transition to the phase τ_2 takes place for a smaller increase in N_3 at the end of the phase τ_1.

(b) But if N_3 is increased in the phase τ_2, then the duration of this phase increases by virtue of the decrease in N_1 and the subsequent decrease in N_2.

(c) Suppose that N_3 is decreased in the phase τ_2. Then N_1 is increased according to (6.1), and this leads to an increase in N_2. The latter causes the phase τ_2 to be shortened. But if N_3 is decreased in the phase τ_1, then its duration increases in view of the foregoing.

(d) According to (6.4), r_3 increases as K_3 is made larger. This implies that the roots ω_1 and ω_2 of equation (6.7) coalesce and disappear. This leads to a quenching of the oscillations.

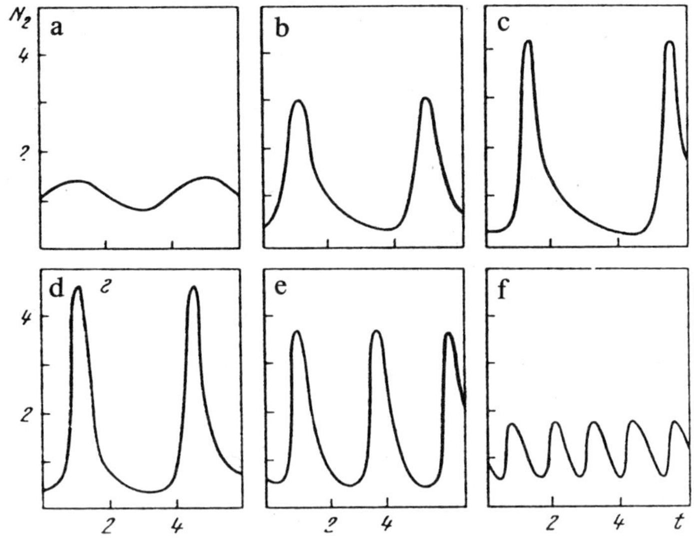

FIGURE 6.1

(e) Something analogous to the preceding happens upon increasing K_2, and hence also r_2, in view of (6.4).

(f) Suppose that N_2 is increased in the phase τ_1. According to (6.3), N_3 then increases, and N_1 decreases by virtue of (6.1). From this and (6.2) it follows that the phase τ_1 is shortened.

Thus, the statements formulated above are identical to those at the beginning of the subsection. The mathematical model (6.1)–(6.3) has thereby tested favorably.

6.2. A relaxation formulation of the problem, and construction of an approximating relay system. In (6.1)–(6.3) we make the normalization substitutions

$$N_s \to K_s N_s, \quad s = 1, 2, 3,$$

and pass to the system

$$\dot{N}_1 = r_1[1 + a(1 - N_3) - N_1]N_1, \tag{6.9}$$

$$\dot{N}_2 = r_2[N_1 - N_2]N_2, \tag{6.10}$$

$$\dot{N}_3 = r_3[N_2 - N_3]N_3 \tag{6.11}$$

for the densities of the substances in the reaction. Below we assume that the parameter a is sufficiently large, and the parameters r_1, r_2, and r_3 are fixed in some way. Theorem 6.1 suggests that under these conditions there must exist stable auto-oscillations for the system (6.9)–(6.11). A numerical computation not only confirms this but also allows us to see that the relaxation cycle constructed as follows is an attractor of the system (6.9)–(6.11): the oscillations in the bromate density are δ-shaped; the oscillations in the cerium density are large, but smoother; the oscillations in the density of bromide have relatively mild slopes.

Thus, on a heuristic level the dynamic behavior of the system (6.9)–(6.11) is clear, and, as in the case of the Hutchinson equation (2.1), its limit object as $a \to \infty$ is a relaxation cycle. However, we encounter substantial difficulties when we try to work out a rigorous theory. This has mainly to do with the fact that the nonzero characteristic exponents of the variational system of the relaxation cycle have finite limits as $a \to \infty$, and its multipliers converge to zero due to the asymptotically large period of the oscillations. In connection with these circumstances, we confine ourselves below to a presentation of a formalism for constructing the so-called zero asymptotics of the relaxation cycle, for the justification of which we refer to numerical integration results for the system (6.9)–(6.11). In particular, this course of action makes the above properties of the characteristic exponents and multipliers of the variational systems intuitively clear.

An even more complicated problem will be solved theoretically in Chapter II. Here, though, we prefer to manage with intuitive ideas, mainly since the system (6.9)–(6.11) does not contain a delay, and hence is not one of our principal objects of study. It is appropriate to mention it, however, because the significance of the system (6.9)–(6.11) in biophysics seems roughly the same as the role of the Hutchinson equation (1.3) in mathematical ecology.

We return to an analysis of the system (6.9)–(6.11). Let us begin by recalling that the ordinary relaxation system

$$\varepsilon\dot{x} = f(x, y), \quad \dot{y} = g(x, y), \qquad x \in R^n, \ y \in R^m, \tag{6.12}$$

is in a certain generality of position a smooth perturbation of the relay system

$$\dot{y} = g(x, y), \qquad x = \begin{cases} \Phi_1(y), \\ \Phi_2(y), \end{cases}$$

where the "towel" $x = \Phi_s(y)$, $s = 1, 2$, is found from the algebraic equation $f(x, y) = 0$. It seems likely that an arbitrary relaxation system described by ordinary differential equations is based on a suitably constructed relay system, which, in the case of the system (6.9)–(6.11) of interest to us, must be taken to depend on a. Our goal is to make a reasonable choice of a relay system approximating (6.9)–(6.11), in order to get some valid information about the asymptotic properties of the relaxation cycle we are interested in.

Thus, in the case of the nontraditional system (6.9)–(6.11) the situation differs strongly from the classical case of the system (6.12), where the zero asymptotics of the relaxation cycle, which is independent of ε, is determined from the relay system (6.13).

The properties mentioned earlier of the relaxation cycle of the system (6.9)–(6.11) suggest that its approximating relay system should be determined as follows. We discard equation (6.9), assuming that in (6.4)

$$N_1 = F(N_3). \tag{6.14}$$

Here

$$F(N_3) = \begin{cases} 0 & \text{for } N_3 > 1, \\ R & \text{for } N_3 < d, \\ 0 \text{ or } R & \text{for } d < N_3 < 1, \end{cases} \tag{6.15}$$

where the large parameter R is equal to the maximal value of the first coordinate of the relaxation cycle of the system (6.9)–(6.11), and $d = d(R) \to 0$ as $R \to \infty$. We describe how the relay system (6.10), (6.11), (6.14) operates. Suppose for definiteness that $N_3(0) > 1$. Then initially $N_1(t) \equiv 0$ in equation (6.10), and therefore $N_2(t)$ is decreasing, which leads to a decrease in $N_3(t)$ with time. There is thus a first moment of time t_0 such that $N_3(t_0) = d$. If the normal switching condition $\dot{N}_3(t_0) < 0$ holds, then the switching takes place: $N_1(t) \equiv R$ at subsequent moments of time in (6.10). For $t > t_0$ the function $N_2(t)$ increases, which leads to an increase in $N_3(t)$. Let t_1 be the first moment of time when $N_3(t_1) = 1$, and the normal switching condition $\dot{N}_3(t_1) > 0$ holds. Then again $N_1(t) \equiv 0$ at subsequent moments of time in (6.10), and so on.

In what follows we construct the asymptotics of a cycle of the relay system (6.10), (6.11), (6.14) with the help of the fitting method. We describe its essence. Arbitrarily, we take two positive numbers τ_1 and τ_2 and set

$$N_1(t) = \begin{cases} R & \text{for } 0 < t < \tau_1, \\ 0 & \text{for } \tau_1 < t < \tau_1 + \tau_2, \end{cases} \tag{6.16}$$

in correspondence with (6.15), and we extend $N_1(t)$ to the remaining values of t by periodicity with period $T = \tau_1 + \tau_2$. Under the condition (6.16) we find the T-periodic function $N_2(t)$ uniquely from (6.10). We substitute it in (6.11) and solve this equation under the initial condition $N_3(0) = d$, requiring that

$$N_3(\tau_1) = 1, \qquad N_3(\tau_1 + \tau_2) = d. \tag{6.17}$$

These equalities, which are called the period equations, allow us to find τ_1 and τ_2 in general. If in addition it turns out that there are no superfluous switchings on a period, that is,

$$N_3(t) < 1 \quad \text{for } 0 < t < \tau_1, \tag{6.18}$$

$$N_3(t) > d \quad \text{for } \tau_1 < t < \tau_1 + \tau_2, \tag{6.19}$$

and the normal switching conditions

$$\dot{N}_3(\tau_1) > 0 \quad \text{and} \quad \dot{N}_3(\tau_1 + \tau_2) < 0 \tag{6.20}$$

hold, then the T-periodic regime constructed in this way is a cycle of the relay system (6.10), (6.11), (6.14).

It follows from the structure of the system (6.9)–(6.11) that the coordinates of its cycle have averages equal to 1. Therefore, we require in addition that

$$R\tau_1 = \tau_1 + \tau_2. \tag{6.21}$$

This restriction enables us to find the functional dependence of d on R in the general case.

We now dwell on the complicated question of the functional dependence of R on a. We assume everywhere below that

$$R = a[1 - c(a)/\ln a], \tag{6.22}$$

where the positive function $c(a)$ is bounded as $a \to \infty$. We later describe a way of making it more precise. First, however, various characteristics of a cycle of the relay system (6.10), (6.11), (6.14) will be determined with an accuracy not affected by the arbitrariness in the choice of $c(a)$.

6.3. Asymptotics of a cycle of the approximating relay system. We let $u = 1/N_2$ and derive from (6.10) that

$$u(t) = \begin{cases} u(0)E(t) + \frac{1}{R}[1 - E(t)], & 0 \le t \le \tau_1, \\ u(\tau_2) + r_2(t - \tau_1), & \tau_1 \le t \le \tau_1 + \tau_2, \end{cases} \tag{6.23}$$

where $E(t) = \exp(-r_2 R t)$ and, by periodicity,

$$u(0) = 1/R + r_2\tau_2/[1 - E(\tau_1)]. \tag{6.24}$$

Let $v = 1/N_3$, $v(0) = 1/d$, and $v(\tau_1) = 1$. Then by (6.11),

$$v(t) = \begin{cases} \frac{1}{d}R_-(t, 0) + r_3 \int_0^t R_-(t, s)\,ds, & 0 \le t \le \tau_1, \\ R_-(t, \tau_1) + r_3 \int_{\tau_1}^t R_-(t, s)\,ds, & \tau_1 \le t \le \tau_1 + \tau_2, \end{cases} \tag{6.25}$$

where

$$R_\pm(t, s) = \exp\left[\pm r_3 \int_s^t \frac{d\sigma}{u(\sigma)}\right].$$

We note at once that for $0 \le t \le \tau_1$

$$R_+(t, 0) = \left[1 - \frac{1}{Ru(0)} + \frac{E(-t)}{Ru(0)}\right]^\gamma, \tag{6.26}$$

where $\gamma = r_3/r_2$, and for $\tau_1 \le s \le t \le \tau_1 + \tau_2$

$$R_-(t, s) = \left[\frac{u(\tau_1) + r_2(s - \tau_1)}{u(\tau_1) + r_2(t - \tau_1)}\right]^\gamma. \tag{6.27}$$

It follows from the continuity condition for the function (6.25) at τ_1 that

$$\frac{1}{d} = R_+(\tau_1, 0) - r_3 \int_0^{\tau_1} R_+(s, 0)\,ds. \tag{6.28}$$

And since this function (6.25) must be $1/d$ at $\tau_1 + \tau_2$, we have, in addition to (6.28),

$$\frac{1}{d} = R_-(\tau_1 + \tau_2, \tau_1) + r_3 \int_{\tau_1}^{\tau_1+\tau_2} R_-(\tau_1 + \tau_2, s)\,ds. \tag{6.29}$$

We equate the right-hand sides of (6.28) and (6.29), and then eliminate τ_1 with the help of (6.21). The result is an equation for determining τ_2, from which it follows in a relatively simple way with the use of (6.26) and (6.27) that

$$\begin{gathered} \tau_2 = (1/r_2)\ln A + O(1/\ln a), \\ A = \begin{cases} A_1 & \text{for } \gamma \le 1, \\ A_2 & \text{for } \gamma > 1, \end{cases} \\ A_1 = a(\ln a)^{(1+1/\gamma)(1+1/\ln a)}/(1 + 1/\gamma)^{(1+1/\ln a)/\gamma}, \\ A_2 = A_1 - (1 + 1/\gamma)^{1/\gamma}/(\ln a)^{1/\gamma}. \end{gathered} \tag{6.30}$$

After this, the asymptotics of τ_1 and d are constructed with the help of (6.21), (6.22), and (6.28).

Let $\underline{N}_2$ and $\overline{N}_2$ be the minimal and maximal values of $N_2(t)$. Then

$$\underline{N}_2 = 1/\ln A + O(1/(\ln a)^3), \tag{6.31}$$

$$\frac{\overline{N}_2}{a} = 1 - \frac{(1 + 1/\gamma)^{1/\gamma}\ln[a(\ln a)^{1+1/\gamma}]}{(\ln a)^{1+1/\gamma}} + O\left(\frac{1}{(\ln a)^{1+1/\gamma}}\right). \tag{6.32}$$

This follows from the fact that, by (6.23), the function $u(t)$ takes its maximal value at zero and its minimum at τ_1.

Finding $\underline{N}_3$ and $\overline{N}_3$ is more difficult, since the behavior of $v(t)$ is more complicated. Its maximum, which is attained at a moment of time asymptotically close to zero, is simpler to determine. Here the simple computations are based on the Taylor representation of $v(t)$. Omitting them, we present the final result at once:

$$\underline{N}_3 = (1 + 1/\gamma)/\ln A + O(1/(\ln a)^3). \tag{6.33}$$

The minimum of $v(t)$ is realized for some t in the interval $(\tau_1, \tau_1 + \tau_2)$. Using (6.25) and (6.27), we see that it coincides with the minimum for $0 < x < 1$ of the function

$$(1 - \Delta)x^\gamma + \Delta/x, \qquad \Delta = u(\tau_1)/(1 + 1/\gamma).$$

Therefore,

$$\frac{\overline{N}_3}{a^{\gamma/(1+\gamma)}} = \frac{1}{(1+\gamma)^{1/(1+\gamma)}} + O\left(\frac{1}{(\ln a)^{1/\gamma}}\right). \tag{6.34}$$

We summarize these constructions as the following proposition.

LEMMA 6.1. *The periodic functions* $N_2(t)$ *and* $N_3(t)$ *are components of a cycle of the relay system* (6.10), (6.11), (6.14).

For a proof we must see that the inequalities (6.18) and (6.19) hold together with the normal switching conditions (6.20); this is elementary.

As a supplement to Lemma 6.1 we note that the cycle we have constructed is orbitally exponentially stable, and its nonzero characteristic exponent is finite.

6.4. On certain properties of solutions of the original system. Let $N_1(0) = 1$, and let $N_2(0)$ and $N_3(0)$ be chosen with (6.31) and (6.33) taken into account. In asymptotic integration of the system (6.9)–(6.11) with the indicated initial conditions the main difficulty involves dealing with the asymptotically small time portion adjacent to zero when there is an essential change in $N_1(t)$. It is rather obvious that the integral of $N_1(t)$ is finite over the interval of its upward flight beginning at once and continuing for a time interval of order $\ln a/r_1 a$. Then $N_1(t)$ is roughly equal to a for a time of order $\ln a/r_2 a$. At the end of this interval there is a sharp increase in $N_3(t)$, which results in $N_1(t)$ falling to very small values. Asymptotic integration of the system (6.9)–(6.11) does not cause problems for the remaining values of t.

This description of the character of behavior of solutions of the system (6.9)–(6.11) with the initial conditions of greatest interest for us implies that the maximal value $\overline{N}_1$ of the component $N_1(t)$ of its relaxation cycle must with high accuracy be attained at a moment of time when the value of the component $N_3(t)$ is minimal. Therefore, it follows from (6.9) and (6.33) that

$$\overline{N}_1 = 1 - \frac{1 + 1/\gamma}{\ln A} + O\left(\frac{1}{(\ln a)^3}\right). \tag{6.35}$$

Using (6.35), we can make the function $c(a)$ in (6.22) more precise, which enables us to compute more precisely the parameters of the relaxation cycle of the relay system (6.10), (6.11), (6.14).

Now a brief word about the C-theorem for the system (6.9)–(6.11). It holds, of course, and its proof is an extension of the description presented above, with use of the technique developed in the preceding subsection. The theorem on C^1-convergence also holds here, but its proof is very cumbersome.

In conclusion we present a table in which the lower rows contain values found from (6.30)–(6.35), and the upper rows contain results of numerical integration of the system (6.9)–(6.11) for the typical case $r_1 = r_2 = r_3 = 1$.

a	T	$\overline{N}_1/a$	$\overline{N}_2/a$	$\overline{N}_3/\sqrt{a}$	$\underline{N}_2$	$\underline{N}_3$
200	8.7	0.76	0.46	0.62	0.116	0.232
	8.4	0.76	0.38	0.71	0.118	0.236
500	9.9	0.79	0.50	0.64	0.101	0.202
	9.6	0.79	0.49	0.71	0.104	0.208
700	10.4	0.80	0.51	0.65	0.096	0.193
	10.1	0.80	0.52	0.71	0.099	0.198
1000	10.8	0.81	0.51	0.66	0.092	0.185
	10.5	0.81	0.55	0.71	0.095	0.190

6.5. Biological and physical systems of reaction-diffusion type. For a whole series of problems in biophysics and mathematical ecology it is typical for the unique attractor of the point model

$$\dot{u} = F(u) \tag{6.36}$$

to be an orbitally exponentially stable cycle $u_0(t)$ of it. In (6.36) $u \in R^n$, $n \geq 2$, and F is a smooth vector-valued function. For an example we refer to the Hutchinson equation (1.3) and the mathematical model of the Belousov reaction (6.9)–(6.11) in the most important range of variation of its parameters. Furthermore, the presence or absence of a delay in the questions considered below does not play a role.

The problem of the adequacy of the point model (6.36) in the cases when experimental material is insufficient, or when its interpretation is ambiguous, is fundamental and difficult. Therefore, it is important to work out general and objective estimations of the quality of mathematical models in biophysics and mathematical ecology. We approach this question from the following aspect. Assume that in some concrete problem it is reasonable (moreover, has great significance) to take diffusion factors into account. It is therefore expedient to consider together with (6.36) the system of reaction-diffusion type

$$\frac{\partial u}{\partial t} = D\Delta u + F(u), \qquad \left.\frac{\partial u}{\partial \nu}\right|_{\partial\Omega} = 0, \tag{6.37}$$

where $D = \operatorname{diag}\{d_j\}$, $d_j > 0$, $j = 1, \dots, n$, and Ω is a bounded domain with piecewise smooth boundary in R^2 or R^3. We regard $C(\Omega)$ as its phase space (space of initial conditions).

Let us begin with the following concept.

DEFINITION 1. The boundary value problem (6.37) is a *simple physical problem* if its homogeneous cycle $u_0(t)$ is stable under any change of the diffusion coefficients.

A sharply focused analysis shows that the existence of stable spatially heterogeneous regimes in simple physical systems depends entirely on good geometric properties of the domain Ω. For example, traveling waves are typical on the circle for this type of system [23], while traveling waves are impossible for the Hutchinson equation with diffusion [41]. It also turns out that most of the more commonly-used nominal mathematical models in biophysics and mathematical ecology are simple physical models [46]. The nature of this phenomenon is simple: all these models are versions of the van der Pol system to one or another extent.

To formulate the next definition we first present facts relating to stability of a homogeneous cycle of the system (6.9)–(6.11) with diffusion factors taken into account. This situation was investigated in [27]. It turns out that in an environment with diffusion a cycle of the system (6.9)–(6.11) is unstable if, first, it has already acquired relaxation features in its form and if, second, a hierarchy of sufficiently small diffusion coefficients is observed,

$$d_1 < d_2 < d_3, \tag{6.38}$$

which correspond to N_1, N_2, and N_3. If under the inequalities (6.38) we begin to decrease the diffusion coefficients proportionally, then the number of modes on which instability is observed is relatively small, and they are displaced in the direction of higher modes. The latter means that stability is restored as the diffusion coefficients decrease on lower modes, while diffusion instability is observed due to modes with higher and higher indices.

DEFINITION 2. The boundary value problem (6.37) is a *complex physical problem* if the mechanism of diffusion instability of its homogeneous cycle $u_0(t)$ bears the character described above.

As of now, we know only two truly substantive examples of complex physical systems of the reaction-diffusion type. One of them is, of course, the mathematical model of the Belousov reaction. The second will be discussed in the next chapter.

It follows from (3.10) that the situation is different in the case of the Hutchinson equation with diffusion: the number of modes on which a Hutchinson cycle is unstable grows monotonically as the diffusion coefficient decreases.

DEFINITION 3. The boundary value problem (6.37) is a *biological problem* if the diffusion instability of its homogeneous cycle $u_0(t)$ bears a Hutchinson character, that is, if the picture described above is obtained upon going into the domain of instability by means of a proportional decrease of the diffusion coefficients.

It seems likely that self-organization regimes arise only in biological systems of reaction-diffusion type.

Of course, this classification of boundary value problems of the form (6.37) must be made more precise and comprehensive as information accumulates about the character and nature of spatially heterogeneous regimes in environments with diffusion.

CHAPTER II

Relaxation Cycles of Finite Amplitude in a Mathematical Model of the Dynamics of Insect Populations

7. A description of the mathematical model, and an existence theorem for relaxation cycles

7.1. A mathematical model for an isolated insect population. The life cycle of insects is complex. For example, we describe briefly the yearly life cycle of *Operophtheria boumata* [5]. Its butterflies emerge from the ground under oak trees in November and December. The wingless females crawl to the trees and climb along the trunks. Winged males gather on the lower part of the trunks. Here they mate with the females, who continue to climb in order to deposit eggs in the cracks in the bark and in the lichen high over the ground. On the average, each female deposits about 150 eggs. The larvae appear at the beginning of April, when the leaf-buds on the oaks open. By the second half of May they finish feeding and descend from the trees on silk threads, bury themselves in the ground, spin cocoons, and pupate. Then the process repeats.

The method presented in §1.2 for modelling biological communities suggests that to describe the dynamics of the variation in number during the active insect phases with complete metamorphosis (the egg and pupa stages are peculiar forms of pregnancy) it is expedient to take the system of differential-difference equations [39]

$$\dot{N}_1 = r_1[1 - a(1 - N_2(t - h_1)/K_2) - N_1(t - h_2)/K_1]N_1, \tag{7.1}$$

$$\dot{N}_2 = r_2[N_1(t - (1 - h_1 - h_2))/K_1 - N_2/K_2]N_2. \tag{7.2}$$

Here N_1 is the number of imagoes, N_2 is the number of larvae, h_1 is the time between the appearances of the larvae and the imagoes, $(1 - h_1)$ is the time between the appearances of the imagoes and the larvae, h_2 is the average lifetime in a year for the population of imagoes, K_1 and K_2 are their average numbers, and $r_1(1 - a)$ is the Malthusian coefficient of linear growth. The parameter r_2 depends on the rate of hatching of the larvae, and hence is not small. The parameter $a < 1$ characterizes the fraction of the larvae that reach the imago stage.

Thus, in composing the system (7.1), (7.2) it is assumed that the dynamical behavior of insect populations is determined mainly by its active phases—the imagoes and the larvae. These phases play a key role in the life cycle of insects: the imago implements the reproduction of the species, and an intraspecies struggle ensues at the larval stage. The very structure of equations (7.1), (7.2) is really a reflection of the biological meaning of these two phases. In other words, we regard it as likely that the passive egg and pupa phases should not enter into the dynamical equations. It seems that the determination of oscillations in numbers during these phases is a secondary problem, which is to some extent analogous to the problem of determining oscillations in the numbers of members of age groups in mammal and fish populations.

Let $r_0 = r_1(1-a)$. It follows from the biological meaning of the Malthusian coefficient of linear growth (cf. (1.4)) that

$$r_0 = \ln pm, \tag{7.3}$$

where p is the fraction of females in the imago population, and m is the average number of eggs deposited by a single female. The determination of the parameters h_1 and h_2 from an analysis of experimental material also does not cause difficulties. As already noted above, the parameter r_2 is relatively large. Therefore, it is reasonable to let r_2 tend to infinity in (7.2), introducing the limit equation

$$\dot{N} = r_0\left[1 + \frac{a}{\varepsilon}N(t-(H_0-h_0)) - \frac{1}{\varepsilon}N(t-h_0)\right]N. \tag{7.4}$$

Here

$$N = N_1/K_1, \quad h_0 = h_2, \quad \varepsilon = 1-a, \quad N_2/K_2 = N(t-(H_0-h_1-h_2)).$$

For this way of deriving (7.4) from (7.1), (7.2) we have $H_0 = 1$. However, it is convenient in what follows to take the parameter H_0 to be an arbitrary integer, in order to encompass cases in which the life cycle of the insects is greater than one year. For example, in a population of the Siberian silkworm the life cycle lasts sometimes one year, and sometimes two years, moreover, with equal frequency (see [4] and [52]). In this case the first term in the square brackets on the right-hand side of (7.2) is replaced by the expression

$$\frac{N_1(t-(1-h_1-h_2)) + N_1(t-(2-h_1-h_2))}{2K_1}.$$

Correspondingly, the term $\frac{a}{\varepsilon}N(t-(H_0-h_0))$ in (7.4) must be replaced by

$$\frac{a}{2\varepsilon}(N(t-(1-h_0)) + N(t-(2-h_0))).$$

We mention also that for the Siberian silkworm

$$r_0 = 5, \qquad h_1 = 0.84, \qquad h_2 = 0.12,$$

the parameter a is probably a little larger than 0.5, and r_2 ranges between 10 and 15.

The system (7.1), (7.2) has by now been rather thoroughly analyzed with the help of numerical methods. We briefly go through the basic properties of the attractors of its solutions with positive initial conditions.

First, it is dissipative. Second, its steady-state regimes bear a well-pronounced δ-shaped character. Third, the biological period (the time between successive spikes in the number or between certain groups of spikes) is very close to 1. Fourth, the stationary regimes often bear a complex character, and a gradual growth in the oscillations is observed beginning with the next phase of decline, that is, the phenomenon of bursts in number, well known in nature, is nicely modelled. Fifth, the stationary regimes are sensitive even to small changes in the parameters. Sixth, a large number of qualitatively different stationary regimes can coexist at the same time.

This list can be continued even further. But it follows from the above properties of solutions of (7.1), (7.2) that difficulties of a technical nature involving the usual indeterminacy of some of the parameters necessarily arise in a fine comparison of theoretical and experimental data. Therefore, the paper [59], which gives a theoretical explanation of classical experiments of Nicholson, Birch, and Crombie (described in [5]), seems not at all simple. Its results, in particular, allow us to reach a favorable conclusion about the biological law expressed in brief form by the system (7.1), (7.2).

We emphasize the following two circumstances in connection with the experimental verification of the model (7.1), (7.2). First, the specific nature of most laboratory experiments is such that the parameter a is relatively small because of the hard competition among the larvae. Second, in modelling them one should more carefully take into account the age structure of the population under consideration (a rather simple version is used in [59], and this adds value to that paper).

From the dynamical point of view the single equation (7.4) is equivalent to the system (7.1), (7.2) when r_2 is not small. However, the stationary regimes of equation (7.4) are much more sensitive to variation of the parameters; that is, the situation is the usual one—either simplicity, or stability. Nevertheless, we prefer the simpler equation (7.4) in what follows, assuming that the parameter ε on its right-hand side is sufficiently small (that is, the automatic regulation system generated by it is energetically overexcited). Under this condition we pose the questions of the existence, asymptotics, and stability of relaxation cycles of finite amplitude for equation (7.4). This problem (posed by the second author as far back as 1982) is important in that its solution enables us to reveal analytically properties of its stationary regimes analogous to properties obtained numerically. In order to clear up also the limits of applicability of the theoretical results, we summarize in a table some important numerical characteristics (obtained by analysis of experimental material in [1], [6], [9], [51], and [67]) of abundant insect species.

Species	r_0	h_0
Operophtheria boumata	3.2	0.125
Bupalus piniarius	3.5	0.09
Colorado [potato] beetle	3.5	0.14
Lasiocampida	3.6	0.125
Eurogaster integriceps	3.7	0.1
European pine sawfly	4	0.09
Panolis flammia	4.5	0.06

7.2. A mathematical model of the parasite-host problem. We assume that a parasite deposits eggs in the larvae of a host, inside which the parasite larvae then develop. From this and from the general principles presented earlier it follows that for the active phases of these insects we have the system of equations [36]

$$\dot{N}_1 = r_1 \left[1 - a_1\left(1 - \frac{N_2(t-h_1)}{K_2}\right) + a_2\left(1 - \frac{M_2(t-(q_1+q_2-\gamma))}{Q_2}\right) - \frac{N_1(t-h_2)}{K_1}\right] N_1, \tag{7.5}$$

$$\dot{N}_2 = r_2 \left[\frac{N_1(t-(1-h_1-h_2))}{K_1} - \frac{N_2}{K_2}\right] N_2, \tag{7.6}$$

$$\dot{M}_1 = p_1 \left[\frac{N_2(t-(h_1+h_2+\gamma))}{K_2} - a_3\left(1 - \frac{M_2(t-q_1)}{Q_2}\right) - \frac{M_1(t-q_2)}{Q_1}\right] M_1, \tag{7.7}$$

$$\dot{M}_2 = p_2 \left[\frac{M_1(t-(1-q_1-q_2))}{Q_1} - \frac{M_2}{Q_2}\right] M_2. \tag{7.8}$$

Here γ is the time between the termination of the flight of the host imago in toto and the parasite imago; $r_1(1-a_1+a_2)$ and $p_1(1-a_3)$ are the Malthusian coefficients of linear growth for the host and parasite, respectively; the parameters a_1 and a_3, each less than 1, characterize the depth of the connection between the larvae and imago populations for the host and parasite, respectively; the parameter a_2 is the coefficient of the pressure of the parasite on the host, which leads to a decrease in the average number of host imagoes by a factor of $(1+a_2)$; the biological meaning of the remaining parameters is clear.

We explain the method of choosing the sizes of the delays in the external feedback chains. Let s_1 be the time between the appearance of parasite larvae and the termination of the flight of the host imago, and let s_2 be the time between the termination of the flight of the parasite imago in the current year and the appearance of the host larvae in the past year. Then we have the equalities

$$s_1 = q_1 + q_2 - \gamma, \qquad s_2 = h_1 + h_2 + \gamma,$$

which make the question more understandable.

We also underscore that M_2 appears on the right-hand side of equation (7.5), and not M_1. This has to do with the fact that the number of host imagoes depends on the number of larvae uncontaminated by the parasite; that is, ultimately on the number of parasite larvae in the corresponding time interval.

As mentioned in the preceding subsection, the coefficients r_2 and p_2 are not small. Letting each go to infinity, we arrive at the markedly simpler system

$$\dot{N} = r[1 - a_1(1 - N(t-(1-h))) + a_2(1 - M(t-(1-\gamma)) - N(t-h)]N, \tag{7.9}$$

$$\dot{M} = p[N(t-(1+\gamma)) - a_3(1 - M(t-(1-q))) - M(t-q)]M. \tag{7.10}$$

Here

$$N = N_1/K_1, \quad M = M_1/Q_1, \quad r = r_1, \quad p = p_1, \quad h = h_2, \quad q = q_2.$$

Finally, for $M \equiv 0$ we arrive at equation (7.4), in which $a = a_1/(1+a_2)$. This implies that the pressure of the parasite deepens the connection between the populations of the host larvae and imagoes.

More complex communities of insects can be modelled in an analogous way (see [16], where there is also a comparison with some natural observations). In a theoretical respect our approach is closely related with ideas in [6] and [7], where the variation of the number of insects in time is interpreted as a self-regulating process.

Below we do not touch on the question of relaxation cycles of the system (7.9), (7.10): the goal of this subsection was to draw attention to an interesting unsolved problem.

7.3. An existence theorem for relaxation cycles. It follows from biological considerations that for small values of ε equation (7.4) can have relaxation cycles of the following character: the function N has order 1 (the imago exists) for roughly the time h_0, and then the values of N are very small (there is no imago) for a time close to $H_0 - h_0$. Our immediate goal is to provide a rigorous meaning to this intuitive idea.

It is convenient to unify equation (7.4) by making the normalizing time substitution $t/(H_0 - h_0) \to t$. We arrive at the equation

$$\dot{N} = \lambda\left[1 + \frac{a}{\varepsilon}N(t-1) - \frac{1}{\varepsilon}N(t-h)\right]N, \tag{7.11}$$

where $\lambda = (H_0 + h_0)r_0$ and $h = h_0/(H_0 - h_0)$.

We introduce the one-dimensional mapping

$$z \to 1 + h + z - [\exp\lambda(z+h) - \exp\lambda z]/\lambda, \tag{7.12}$$

the right-hand side of which is denoted by F. This mapping is said to have an exponentially stable cycle

$$z_0, \dots, z_k, \tag{7.13}$$

if the numbers (7.13) are distinct, $F(z_i) = z_{i+1}$, $i = 0, \dots, k$, $z_{k+1} = z_0$, and the multiplier $F'(z_0)\cdots F'(z_k)$ is less than 1 in modulus.

THEOREM 7.1. *Suppose that the one-dimensional mapping* (7.12) *has an exponentially stable cycle* (7.13). *Then for* $h < 1$ *there is an* $\varepsilon_k > 0$ *such that for* $0 < \varepsilon < \varepsilon_k$ *equation* (7.11) *has a relaxation cycle connected in a certain way with the cycle* (7.13) *of the mapping* (7.12), *the general character of which is described above in terms of* (7.4).

We begin our proof with the simplest case when the cycle of the mapping (7.12) is its equilibrium state

$$z_0 = \frac{1}{\lambda}\ln\frac{\lambda(1+h)}{\exp\lambda h - 1}. \tag{7.14}$$

Further, we assume that $\lambda(1+h) < 2$, that is, the equilibrium state (7.14) is exponentially stable.

In the phase space $C(-1, 0)$ of equation (7.11) we consider only nonnegative functions, of course. In this class of functions we distinguish a closed convex set S_0 such that the first-return operator Π is defined on it, and $\Pi S_0 \subset S_0$.

Let

$$\overline{M} = \exp\lambda[z_0 + h + \varepsilon^{1/3}], \qquad \underline{M} = \exp\lambda[z_0 + h - \varepsilon^{1/3}],$$

$$\underline{N}(t, \varepsilon) = \underline{M}\exp\left\{\lambda(t+1) - \frac{\varepsilon}{a\underline{M} + \varepsilon}\left[-1 + \exp\lambda\frac{a\overline{M}+\varepsilon}{\varepsilon}(t+1)\right]\right\}.$$

By definition, $N(t, \varepsilon) \in S_0$ if: first,

$$(7.15) \qquad \underline{N}(t, \varepsilon) \le N(t, \varepsilon) \le \overline{M}, \qquad -1 \le t \le -1 + \delta_1(\varepsilon),$$

where $\delta_1(\varepsilon) = (\frac{\varepsilon}{\lambda} \ln \frac{1}{\varepsilon} + \varepsilon z_0) \exp[-\lambda(z_0 + h)]$; second, $N(t, \varepsilon)$ is monotonically decreasing for

$$(7.16) \qquad -1 + \delta_1(\varepsilon) \le t \le -1 + \delta(\varepsilon), \qquad \delta(\varepsilon) = \delta_1(\varepsilon) + \delta_2(\varepsilon),$$

where $\delta_2(\varepsilon) = (\frac{\varepsilon}{\lambda} \ln \frac{1}{\varepsilon} + \varepsilon(z_0 + h)) \exp[-\lambda z_0 + 3\exp(-\lambda h)]$; third,

$$(7.17) \qquad N(t, \varepsilon) \le \varepsilon \begin{cases} \exp\left[-\frac{m(1-\delta(\varepsilon)+t)}{\varepsilon}\right], & -1 + \delta(\varepsilon) \le t \le -1/2, \\ \exp \frac{m(1-2\delta(\varepsilon)t)}{\varepsilon}, & -1/2 \le t \le 0, \end{cases}$$

where the constant $m > 0$ serves all the functions in S_0; fourth, $N(0, \varepsilon) = \varepsilon$; finally, fifth,

$$(7.18) \qquad z_0 - \varepsilon^{1/2} \le z_{\text{in}} \le z_0 + \varepsilon^{1/2},$$

where

$$(7.19) \qquad z_{\text{in}} = \frac{1}{\varepsilon} \int_{-1}^{-1+\delta(\varepsilon)} N(t, \varepsilon)\, dt - \frac{1}{\lambda} \ln \frac{1}{\varepsilon}.$$

Let us keep the notation $N(t, \varepsilon)$, $t \ge 0$, for a solution of equation (7.11) with initial condition in S_0. We mention some important properties for what follows. The inequalities (7.17) and (7.18) imply that for $h \le t \le h + \delta(\varepsilon)$

$$(7.20) \qquad N(t, \varepsilon) \le \exp \lambda(z_0 + h + 2\varepsilon^{1/2}).$$

An immediate computation shows that

$$(7.21) \qquad \int_{-1}^{-1+\delta_1(\varepsilon)} \underline{N}(t, \varepsilon)\, dt > \frac{\varepsilon}{\lambda} \ln \frac{1}{\varepsilon} + \varepsilon z_0 - \varepsilon \frac{\exp(-\lambda h)}{\lambda}.$$

We derive from (7.17) and (7.21) that, for $\delta_1(\varepsilon) \le t \le h$,

$$(7.22) \qquad N(t, \varepsilon) \ge \exp[\lambda z_0 - 2\exp(-\lambda h)].$$

Using (7.17), (7.20), and (7.22) along with (7.11), we conclude that, for $h + \delta_1(\varepsilon) \le t \le \min(2h, 1) = h_*$,

$$(7.23) \qquad \begin{aligned} N(t, \varepsilon) \le \exp \lambda \Big\{ & z_0 + h + 3\varepsilon^{1/2} + [t - h - \delta_1(\varepsilon)] \\ & \times \left(1 - \frac{1}{\varepsilon} \exp[\lambda z_0 - 2\exp(-\lambda h)]\right) \Big\}. \end{aligned}$$

The inequality (7.23) enables us to conclude that for $h + \delta(\varepsilon) \le t \le h_*$ the solution $N(t, \varepsilon)$ satisfies an inequality of the type (7.17) and that

$$(7.24) \qquad \int_{h+\delta_1(\varepsilon)}^{h_*} N(t, \varepsilon)\, dt \le \varepsilon \exp \lambda[h + 3\exp(-\lambda h)].$$

We denote by t_N the moment of time such that $N(t_N, \varepsilon) = \varepsilon$, where, by (7.22) and (7.23),

$$(7.25) \qquad h < t_N < h + \delta(\varepsilon).$$

The interval $0 \le t \le t_N$ is called the *spike time* of $N(t, \varepsilon)$.

It follows from the structure of equation (7.11) that for $h_* \le t \le 1$ the solution $N(t, \varepsilon)$ also satisfies an inequality of the type (7.17), and its essential growth begins at $t > 1$. It is also clear that the solution $N(t, \varepsilon)$ first attains the value ε at some moment of time T neither later nor earlier than when the previous drop is compensated. Therefore, by (7.18) and (7.25),

$$|T - 1 - h| < \delta(\varepsilon). \tag{7.26}$$

We divide the left-hand and right-hand sides of (7.11) by $N(t, \varepsilon)$ and integrate the result over the interval $[0, T]$. Using (7.17), (7.26), and (7.19), we transform the resulting expression to the form

$$z_{\text{out}} = F(z_{\text{in}}) + O(\varepsilon \ln(1/\varepsilon)), \tag{7.27}$$

where the indicated order of smallness is uniform with respect to the choice of the initial condition in S_0, and

$$z_{\text{out}} = \frac{1}{\varepsilon} \int_{T-1}^{t_N} N(t, \varepsilon)\, dt - \frac{1}{\lambda} \ln \frac{1}{\varepsilon}. \tag{7.28}$$

It follows from (7.18), the conditions of the theorem, and (7.27) that

$$z_0 - q\varepsilon^{1/2} \le z_{\text{out}} \le z_0 + q\varepsilon^{1/2}, \qquad 0 < q < 1. \tag{7.29}$$

It follows from the left-hand inequality in (7.18), the right-hand inequality in (7.15), and (7.11) that for $h \le t \le h + \delta_1(\varepsilon)$

$$N(t, \varepsilon) \ge \exp\left\{\lambda(z_0 - 2\varepsilon^{1/2} + t) - \frac{\varepsilon}{a\overline{M} + \varepsilon}\left[-1 + \exp \lambda \frac{a\overline{M} + \varepsilon}{\varepsilon}(t - h)\right]\right\}. \tag{7.30}$$

From this we derive the following analogue of (7.21):

$$\int_h^{h+\delta_1(\varepsilon)} N(t, \varepsilon)\, dt > \frac{\varepsilon}{\lambda} \ln \frac{1}{\varepsilon} + \varepsilon z_0 - \varepsilon \frac{\exp(-\lambda h)}{\lambda}. \tag{7.31}$$

Assume that

$$T = 1 + h + \varepsilon\gamma \exp[-\lambda(z_0 + h)], \tag{7.32}$$

where $\gamma > 0$ is such that (7.26) holds. Under the condition (7.32) we derive from (7.20) and (7.24) that

$$\begin{aligned} \int_{T-1}^{t_N} N(t, \varepsilon)\, dt &\le \int_{h+\varepsilon\gamma \exp[-\lambda(z_0+h)]}^{h_*} N(t, \varepsilon)\, dt \\ &\le [\delta_1(\varepsilon) - \varepsilon\gamma \exp(-\lambda(z_0 + h))] \exp(z_0 + h + 2\varepsilon^{1/2}) \\ &\quad + \varepsilon \exp \lambda[h + 3\exp(-\lambda h)] \\ &\le \frac{\varepsilon}{\lambda} \ln \frac{1}{\varepsilon} + \varepsilon(z_0 - \gamma) + \varepsilon \exp \lambda[h + 4\exp(-\lambda h)]. \end{aligned} \tag{7.33}$$

We conclude from (7.28), (7.29), and (7.33) that, necessarily,

$$\gamma \le \exp \lambda[h + 5\exp(-\lambda h)]. \tag{7.34}$$

Assume that

$$T = 1 + h - \varepsilon\gamma \exp[-\lambda(z_0 + h)]. \tag{7.35}$$

Since the solution $N(t, \varepsilon)$ differs from $\exp \lambda(z_0 + t)$ by a quantity of order $\varepsilon^{1/2}$ uniformly with respect to the initial condition if $h - \varepsilon\gamma \exp[-\lambda(z_0 + h)] \le t \le h$, we have in view of (7.31)

$$\int_{T-1}^{t_N} N(t, \varepsilon)\, dt \ge \frac{1}{2}\varepsilon\gamma + \frac{\varepsilon}{\lambda} \ln \frac{1}{\varepsilon} + \varepsilon z_0 - \varepsilon \frac{\exp(-\lambda h)}{\lambda}. \tag{7.36}$$

From a comparison of (7.36) with the right-hand inequality in (7.29) we deduce that

$$\gamma \le 3\exp(-\lambda h)/\lambda. \tag{7.37}$$

The inequalities (7.34) and (7.37) lead to the following estimate, which is subtler than (7.26):

$$|T - 1 - h| \le \varepsilon h m_0, \tag{7.38}$$

where the constant m_0 is bounded as $h \to 0$. In particular, it follows from (7.38) that the period of the desired cycle of (7.4) is close to H_0 with good accuracy.

We remark that in view of (7.30) and the asymptotic property just noted for $N(t, \varepsilon)$ the quantities

$$\int_h^{(1+\varepsilon m_0)h} N(t, \varepsilon)\, dt, \quad \int_{(1-em_0)h}^{h} N(t, \varepsilon)\, dt$$

have order ε uniformly with respect to the initial condition. From this and the inequalities (7.20) and (7.30) it follows that $N(T + t, \varepsilon)$ satisfies (7.15) (we recall that, in particular, $\varepsilon^{1/2}$ is replaced by $\varepsilon^{1/3}$ at the same time, and this enables us to compensate for the difference between T and $1 + h$). The monotone decrease of $N(T + t, \varepsilon)$ for values of t satisfying (7.16) is ensured by (7.22) and the structure of the right-hand side of (7.11). The inequality (7.17) for $N(T + t, \varepsilon)$ with

$$m = \frac{\lambda h(1 + h)}{2[1 - \exp(-\lambda h)]} \exp[\lambda h - 2\exp(-\lambda h)]$$

is a consequence of (7.14), (7.22), and the structure of equation (7.11). Thus, the first-return operator $\Pi(N) = N(T+t, \varepsilon)$ is constructed on the set S_0 and transforms this set into itself.

In the general case when the mapping (7.12) has an exponentially stable cycle (7.13) with $k \ge 1$ we can confine ourselves to some explanatory remarks.

The set S_0 is constructed as before, and the first-return operator is represented in the form $\Pi = \Pi_k \cdot \Pi_{k-1} \cdots \Pi_0$. It is then shown that the operator Π_0 transforms S_0 into S_1, where the properties of S_1 are analogous to those of S_0, but with z_0 replaced by z_1. Further, if

$$z_0 - \varepsilon^{1/2} \le z_{\text{in}}^0 \le z_0 + \varepsilon^{1/2},$$

then

$$z_1 - q_0\varepsilon^{1/2} \le z_{\text{in}}^1 \le z_1 + q_0\varepsilon^{1/2},$$

where z_{in}^1 is computed from a formula analogous to (7.28), and q_0 is just a little larger than $|F'(z_0)|$. Continuing this process, we see that Π is defined on S_0 and transforms this set into itself. The proof is complete.

To summarize our proof, we say that the invariant set S_0 has such a complicated structure that to guess it means actually to predict rather nontrivial asymptotic properties of the periodic solutions.

7.4. On the limits of applicability of Theorem 7.1. Setting $M = \exp\lambda(z+h)$, we transform the mapping (7.12) to the form

$$M \to M \exp\{r_0 H_0 - [1 - \exp(-r_0 h_0)]M\}, \tag{7.39}$$

where we have made a transition to the parameters of equation (7.4) at the same time. It follows from the proof of Theorem 7.1 that

$$\varepsilon \ll \min M_i, \tag{7.40}$$

where $M_i = \exp\lambda(z_i + h)$, and the z_i form the cycle (7.13). The restriction (7.40) is not burdensome when the attractor of the mapping (7.39) consists of an exponentially stable equilibrium. A numerical integration of (7.11) carried out by N. B. Fedotov showed that in this case, as the energy parameter a increases, a condensation of trajectories gives rise to a stable cycle that is preserved under any reasonable decrease of ε. For $\lambda(1+h) > 2$ the situation is analogous to start with, but as ε decreases, the cycle undergoes different bifurcations, and not necessarily according to the law given by the mapping (7.12). The reason is simple: the right-hand side of (7.40) becomes extremely small as λ increases. In particular, for most abundant insect species the right-hand side of (7.40) is small, as follows from the numerical characteristics of them given in the table.

7.5. The subsequent scheme of investigation. It follows from the proof of Theorem 7.1 that as $\varepsilon \to 0$ the first-return operator Π converges on S_0 in a rather weak and peculiar sense to the cycle (7.13) of the mapping (7.12). The absence of a limit object in a nice topology is the main cause of the ensuing difficulties.

In the next section we will be able to construct the asymptotics only of the simplest relaxation cycle, the one that corresponds to the exponentially stable equilibrium state (7.14) of the mapping (7.12), under the additional assumption that the delay h in (7.11) is small. We underscore that we can construct the asymptotics of just the cycle, and not of a family of solutions with close initial conditions. Therefore, the theorem on C^1-convergence reduces precisely to a computation of the limit values of the multipliers of the variational equation. We regard it as a nontrivial fact that as $\varepsilon \to 0$ countably many of them have nonzero limits (our not at all simple proof occupies the whole of §9).

At the end of the chapter we turn to a new question: the existence and strong instability of relaxation cycles of finite amplitude for equation (7.11) with a more complicated time ordering.

8. Asymptotics of the simplest relaxation cycle

8.1. The basic boundary value problem. As already mentioned, we are directing our main attention to the construction of the asymptotics of the simplest relaxation cycle, whose period is equal to $1 + h + \varepsilon\nu$ according to the estimate (7.38), where the parameter ν depends on ε and is to be determined. At the end of the section we mention a way of extending our technique to the general case.

Let $N(t, \varepsilon)$ be the periodic solution of equation (7.11) under consideration. We introduce the notation

$$\varphi(t, \varepsilon) = \frac{1}{\varepsilon} N(t, \varepsilon), \qquad -1 + h + \varepsilon\nu \le t \le h + \varepsilon\nu,$$

$$\Phi_\varphi(t, \varepsilon) = \frac{1}{\varepsilon} N(t + 1 + h + \varepsilon\nu, \varepsilon), \qquad -1 \le t \le 0.$$

Obviously, for $-1+h+\varepsilon\nu \leq t \leq 0$ we must have on the cycle the equality

$$\varphi(t, \varepsilon) = \Phi_\varphi(t, \varepsilon). \tag{8.1}$$

Further, it follows from (7.11) that for $0 \leq t \leq h+\varepsilon\nu$

$$\varphi(t, \varepsilon) = \exp\left\{\lambda t + \lambda a \int_{-1}^{t-1} \Phi_\varphi(s, \varepsilon)\, ds - \lambda \int_{-h}^{t-h} \varphi(s, \varepsilon)\, ds\right\}. \tag{8.2}$$

We simplify the central equality (8.2) connected with the spike part of the cycle. This can and must be done, since on the time interval under consideration the principal contribution to the formation of $\varphi(t, \varepsilon)$ is due to its power asymptotics.

It follows from (7.11) that

$$\begin{aligned}\Phi_\varphi(t, \varepsilon) &= \varphi(h+\varepsilon\nu, \varepsilon)\\ &\times \exp\left\{\lambda(t+1) + \frac{\lambda a}{\varepsilon}\int_{-1+h+\varepsilon\nu}^{t+h+\varepsilon\nu} N(s, \varepsilon)\, ds - \frac{\lambda}{\varepsilon}\int_{\varepsilon\nu}^{t+1+\varepsilon\nu} N(s, \varepsilon)\, ds\right\}.\end{aligned} \tag{8.3}$$

From (8.2) and (8.3) we deduce the following equality for $0 \leq t \leq h$, up to quantities of order $\exp(-m\varepsilon)$, $m > 0$:

(8.4)

$$\begin{aligned}\varphi(t, \varepsilon) = \exp\Bigg\{&\lambda t - \lambda\int_{-h}^{t-h} \varphi(s, \varepsilon)\, ds\\ &+ \lambda a \varphi(h+\varepsilon\nu, \varepsilon)\int_{-1}^{t-1} \exp\left[\lambda(s+1) - \lambda\int_{\varepsilon\nu}^{s+1+\varepsilon\nu} \varphi(\sigma, \varepsilon)\, d\sigma\right] ds\Bigg\}.\end{aligned}$$

The quantity $\varphi(h+\varepsilon\nu, \varepsilon)$ appears on the right-hand side of (8.4). It is simple to compute. Indeed, in view of the proof of Theorem 7.1 the equality

$$N(t, \varepsilon) = \exp\lambda(z+t) \tag{8.5}$$

holds with the accuracy noted above for $\delta \leq t \leq h-\delta$ (the positive parameter δ is small but fixed), where $z = z(\varepsilon)$ and $z(0) = z_0$, and

$$N(h+\varepsilon\nu, \varepsilon) = \exp\lambda(z+h+\varepsilon\gamma), \tag{8.6}$$

where the parameter $\gamma = \gamma(\varepsilon)$ has order 1 as $\varepsilon \to 0$. This implies that

$$\varepsilon\varphi(h+\varepsilon\nu, \varepsilon) = \exp\lambda(z+h+\varepsilon\gamma). \tag{8.7}$$

We pass to a derivation of the basic differential equation playing a fundamental role in the question of the asymptotics of a cycle on the spike part. With this goal we differentiate (8.4) with respect to time and use (8.7). We get

$$\dot\varphi = \lambda\left[1 - \varphi(t-h, \varepsilon) + \frac{a}{\varepsilon}\exp\left\{\lambda(z+h+\varepsilon\gamma+t) - \lambda\int_{\varepsilon\nu}^{t+\varepsilon\nu} \varphi(\sigma, \varepsilon)\, d\sigma\right\}\right]\varphi. \tag{8.8}$$

It is sufficiently clear that the main difficulty in constructing the asymptotics of a relaxation cycle has to do with the beginning of its spike. This means that instead of (8.8) we study the simpler equation

$$\dot\varphi = \lambda\left[1 + \frac{a}{\varepsilon}\exp\left\{\lambda(z+h+\varepsilon\gamma+t) - \lambda\int_{\varepsilon\nu}^{t+\varepsilon\nu} \varphi(\sigma, \varepsilon)\, d\sigma\right\}\right]\varphi. \tag{8.9}$$

We give a precise meaning to the concept of a solution of (8.9), assuming the parameter ν to be positive (which will be proved rigorously below). If in (8.9) we set

$$\psi = \int_{\varepsilon\nu}^{t} \varphi(\sigma, \varepsilon)\, d\sigma,$$

then we get the equation

$$\ddot{\psi} = \lambda \left[1 + \frac{a}{\varepsilon} \exp\{\lambda(z + h + \varepsilon\gamma + t) - \lambda\psi(t + \varepsilon\nu, \varepsilon)\}\right] \dot{\psi} \tag{8.10}$$

with an advanced argument. We fix $0 < t_0 < h$ and pose for equation (8.10) the Cauchy problem of finding a solution for $t < t_0$ if it is known that for $t_0 \le t \le t_0 + \varepsilon\nu$

$$\varepsilon\psi = (1/\lambda)[\exp \lambda(z + t) - \exp \lambda z] - c_0 \varepsilon \ln(1/\varepsilon) + \varepsilon c_1, \tag{8.11}$$

$$\varepsilon\dot{\psi} = \exp \lambda(z + t), \tag{8.12}$$

where $c_0 = \lambda^{-1} \exp(-\lambda h)$, and c_1 is an arbitrary parameter which together with ν is determined from the supplementary conditions

$$\psi(\varepsilon\nu, \varepsilon) = 0, \quad \dot{\psi}(0, \varepsilon) = 1. \tag{8.13}$$

We point out that the right-hand side of (8.12) and the principal part of the right-hand side of (8.11) are consequences of (8.5). The necessity of modifying the right-hand side of (8.11) has to do with the equalities (8.13), the second of which follows from (8.2), which we have already mentioned and which is explained below.

It will follow from the constructions below that the basic boundary value problem (8.10)–(8.13) is uniquely solvable. This enables us to construct the asymptotics of a cycle with any desired accuracy. The parameters z and γ in (8.10) are determined in passing. However, we pose a more restricted problem: to construct for a cycle an asymptotic expression that is essential for clearing up its stability properties.

The verification of the asymptotics of a solution of the problem (8.10)–(8.13) is standard and is based on an investigation of the linearized equation (8.10). But one new point arises: because of the singularity of the problem (more details about its nature will be given), the proof of even simple asymptotic formulas requires knowledge of a sufficiently refined asymptotic expression. Therefore, we present algorithms for its computation.

8.2. Algorithms for constructing the asymptotics of a solution of the boundary value problem (8.10)–(8.13) on slow and rapid motion parts. For $\varepsilon^{\lambda_0} \le t \le t_0$, $0 < \lambda_0 < 1$, i.e., on a slow motion part, equation (8.10) implies that the formula (8.11) is preserved for a solution of (8.10)–(8.12) to within $\exp(-m/\varepsilon^{1-\lambda_0})$, $m > 0$.

We make the substitutions $t = \varepsilon\tau$, $\psi(\varepsilon\tau, \varepsilon) = \varepsilon v(\tau, \varepsilon)$ in a neighborhood of the zero moment of time, i.e., on a rapid motion part, obtaining the equation

$$\ddot{v} = [\varepsilon\lambda + a\varkappa \exp\{\varepsilon\lambda(\gamma + \tau) - \varepsilon\lambda v(\tau + \nu, \varepsilon)\}]\dot{v}, \tag{8.14}$$

which by (8.13) should be supplemented by the conditions

$$v(\nu, \varepsilon) = 0, \quad \dot{v}(0, \varepsilon) = 1. \tag{8.15}$$

Here $\varkappa = \lambda \exp \lambda(z + h)$, and z will be regarded as an independent parameter close to z_0.

Setting

$$v(\tau, \varepsilon) = v_0(\tau) + \sum_{k=1}^{\infty} \varepsilon^k v_k(\tau) \tag{8.16}$$

in (8.14), (8.15), where (see (8.14), (8.15) with $\varepsilon = 0$)

$$v_0(\tau) = (1/\varkappa)(\exp \varkappa\tau - \exp \varkappa\nu), \tag{8.17}$$

and equating the coefficients of like powers of ε, we arrive at the following recursion sequence of differential equations for determining the unknown functions v_k, $k \geq 1$:

$$\ddot{v}_k = \varkappa\dot{v}_k + f_k(\tau), \tag{8.18}$$

where f_k depends only on the v_m with indices $m \leq k-1$. Supplementing (8.18) by the conditions

$$v_k(\nu) = 0, \qquad \dot{v}_k(0) = 0, \qquad k \geq 1, \tag{8.19}$$

we determine the coefficients of the series (8.16) successively. For example, for v_1 we have the equation

$$\ddot{v}_1 - \varkappa\dot{v}_1 = (\lambda - \varkappa)\dot{v}_0 + \varkappa\lambda[\gamma + \tau - v_0(\tau + \nu)]\dot{v}_0,$$

from which by (8.19) we find

(8.20)
$$v_1 = R(\tau) - R(\nu),$$

$$R(\tau) = \left[\frac{1 - \lambda\gamma}{\varkappa} + \left(-1 + \lambda\gamma + \frac{\lambda}{\kappa}\exp \varkappa\nu\right)\tau + \frac{\lambda}{2}\tau^2\right]\exp \varkappa\tau - \frac{\lambda}{2\varkappa^2}\exp(2\tau + \nu).$$

The equalities (8.17) and (8.20) together lead us by induction to the representations

$$v_k(\tau) = r_k + \sum_{l=1}^{k+1} P^k_{2(k-l+1)}(\tau)\exp l\varkappa\tau, \qquad k \geq 0, \tag{8.21}$$

where the r_k are constants, and the $P^k_m(\tau)$ are polynomials of degree m. The representations (8.21) will be used in §8.4 in the construction of higher approximations on the turning part.

8.3. The zero approximation of a solution of the boundary value problem (8.10)–(8.13) on the turning part. The investigation of the turning part is connected with the substitutions

$$t = \frac{\varepsilon}{\varkappa}\ln\frac{1}{\varepsilon s}, \qquad w = \psi\left(\frac{\varepsilon}{\varkappa}\ln\frac{1}{\varepsilon s}, \varepsilon\right). \tag{8.22}$$

Making these substitutions in (8.10), we get the equation

$$sw'' + w' = -\left[\varepsilon\frac{\lambda}{\varkappa} + a\exp\left\{\varepsilon\lambda\left(\gamma + \frac{1}{\varkappa}\ln\frac{1}{\varepsilon s}\right) - \lambda w(sq, \varepsilon)\right\}\right]w', \tag{8.23}$$

where $q = \exp(-\varkappa\nu)$, and the prime stands for differentiation with respect to s. If we next set $\varepsilon = 0$ in (8.23), we arrive at the equation

$$sw_0'' + w_0' = -w_0'\exp[-\lambda w_0(sq)] \tag{8.24}$$

for determining the zero approximation $w_0(s)$ of the function $w(s, \varepsilon)$.

Using the substitutions (8.22) in (8.11) and (8.17), we see that we must construct a solution $w_0(s)$ of (8.24) with the asymptotic representations

$$w_0 \sim 1/\varkappa s\,, \quad s \to \infty\,, \qquad w_0 \sim -c_0 \ln s + c_1\,, \quad s \to +0. \tag{8.25}$$

Let us show that for suitable ν and c_1 equation (8.24) has a solution with the properties (8.25).

We supplement (8.24) by the initial conditions

$$sw_0'|_{s\to+0} = -c_0\,, \qquad (w_0 + c_0 \ln s)|_{s\to+0} = c_1. \tag{8.26}$$

Local solvability of the problem (8.24), (8.26) is established as follows. Let

$$w_0 = v - c_0 \ln s\,, \qquad \omega = s^{1-\lambda c_0} v'.$$

Then to determine $\omega(s)$ we get the integral equation

$$\omega(s) = c_0 \left(\frac{q}{s}\right)^{\lambda c_0} \int_0^s \frac{\Omega(\sigma)}{\sigma^{1-\lambda c_0}}\, d\sigma - \left(\frac{q}{s}\right)^{\lambda c_0} \int_0^s \sigma^{2\lambda c^0 - 1} \omega(\sigma)\Omega(\sigma)\, d\sigma \tag{8.27}$$

with a weak singularity, where

$$\Omega(\sigma) = \exp\left[-\lambda c_1 - \lambda q^{\lambda c_0} \int_0^\sigma \frac{\omega(\tau q)}{\tau^{1-\lambda c_0}}\, d\tau\right].$$

We apply the contraction mapping principle to this integral equation in the space $C(0\,, s_0)$ for sufficiently small s_0.

LEMMA 8.1. *The solution of the problem* (8.24), (8.26) *is defined on the whole positive semiaxis,* $w_0'(s) \le 0$, *and it has the asymptotic expansions*

$$w_0 = -c_0 \ln s + c_1 + \sum_{k=1}^{\infty} c_{k+1} s^{k\lambda c_0}\,, \qquad s \to +0\,, \tag{8.28}$$

$$w_0 = x_0 + \frac{x_1}{s^a} + \sum_{k=2}^{\infty} \frac{x_k}{s^{k\alpha}}\,, \qquad s \to \infty\,, \tag{8.29}$$

where $\alpha = \exp(-\lambda x_0)$.

PROOF. We first show that $w_0'(x) \le 0$ for $s \in (0\,, Q)$, where $(0\,, Q)$ is the maximal interval of existence of a solution.

The inequality $w_0'(s) < 0$ is obvious for small s in view of the first condition in (8.26). Suppose that there exist $0 < s_0 < s_1$ such that $w_0'(s) > 0$ for $s_0 < s \le s_1$ and $w_0'(s_0) = 0$. Under these conditions and integrating (8.24) over $[s_0\,, s_1]$, we get

$$0 < s_1 w_0'(s_1) = -\int_{s_0}^{s_1} \exp[-\lambda w_0(sq)] w_0'(s)\, ds < 0\,,$$

which is impossible.

If $Q < \infty$, then, by the proven monotonicity of $w_0(s)$,

$$\lim_{s\to Q-0} w_0(s) = -\infty. \tag{8.30}$$

But then the relations

$$-\int_0^s \exp[-\lambda w_0(\sigma q)] w_0'(\sigma)\, d\sigma = c_0 + s w_0'(s) \le c_0$$

imply convergence of the improper integral

$$-\int_0^Q \exp[-\lambda w_0(\sigma q)]w_0'(\sigma)\,d\sigma. \tag{8.31}$$

On the other hand, (8.30) gives us the existence of an $s_0 \in (0, Q)$ such that $w_0(s) < 0$ for $s \in [s_0, Q)$. And from this and the obvious inequality

$$-\int_{s_0}^Q \exp[-\lambda w_0(\sigma q)]w_0'(\sigma)\,d\sigma \geq -\int_{s_0}^Q w_0'(\sigma)\,d\sigma = \infty$$

we arrive at a contradiction to the convergence of the integral (8.31).

Thus, the function $w_0(s)$ is defined for all $s > 0$. The impossibility of the limit equality (8.30) as $s \to \infty$ is proved similarly (with Q replaced by ∞). The function $w_0(s)$ thus has the finite limit x_0 as $s \to \infty$. Using this in (8.24), we easily verify the expansion (8.29) by the method of undetermined coefficients. The validity of (8.28) is derived similarly from (8.27). The lemma is proved.

It follows from the first condition in (8.25) and the asymptotic equality (8.29) that further progress involves solvability of the system

$$x_0(\nu) = 0, \qquad x_1(\nu, c_1) = 1/\varkappa \tag{8.32}$$

of algebraic equations, where the left-hand side of the first equation does not depend on c_1. This follows from the assertion analogous to Lemma 8.1 but for the equation obtained from (8.24) by the substitution $s\exp(-c_1/c_0) \to s$.

LEMMA 8.2. *There is a positive number $\overline{h} < 1$ such that for $0 < \overline{h} \leq h$ the system (8.32) is uniquely solvable for $z = z_0$, and its solution $(\nu^0(h), c_1^0(h))$ satisfies the asymptotic equalities*

$$\nu^0(h) = 2h^2 + O(h^3), \qquad c_1^0(h) = -\frac{1}{\lambda}\ln\frac{1}{h} + O(h). \tag{8.33}$$

PROOF. By (7.14), $\varkappa|_{z=z_0} = \lambda\varkappa_0(h)/h$, where $\varkappa_0(0) = 1$. Therefore, equation (8.24) formally depends in a singular way on $h \to 0$. However, after the substitutions $\nu = \mu h/\lambda$ (see (7.38)), $s = h\sigma$, $c_1 + c_0\ln(1/h) = \tilde{c}_1$ we get, instead of problem (8.24), (8.26), the Cauchy problem

$$\sigma w_0'' + w_0' = -w_0'\exp[-\lambda w_0(\sigma q_0)], \tag{8.34}$$

$$\sigma w_0'|_{\sigma\to+0} = -c_0, \qquad (w_0 + c_0\ln\sigma)|_{\sigma\to+0} = \tilde{c}_1, \tag{8.35}$$

where $q_0 = \exp[-\mu\varkappa_0(h)]$, whose dependence on μ and h is regular. According to Lemma 8.1, the solution of the problem (8.34), (8.35) has as $\sigma \to \infty$ the asymptotic representation

$$w_0(\sigma) = y_0 + y_1/\sigma^a + O(1/\sigma^{2\alpha}), \qquad \alpha = \exp(-\lambda y_0).$$

Therefore, equations (8.32) are equivalent to the system

$$y_0(\mu, h) = 0, \qquad y_1(\mu, \tilde{c}_1, h) = 1/\lambda\varkappa_0(h). \tag{8.36}$$

It is essential that for $\mu = 0$ equation (8.34) is integrable by quadratures, and the solution of interest to us is supplied by the formula

$$w_0(\sigma) = (1/\lambda)\ln(\exp\lambda\tilde{c}_1/\sigma^\beta + 1/\beta), \qquad \beta = \exp(-\lambda h). \tag{8.37}$$

Therefore, $y_0(0, h) \equiv h$ and

$$\left.\frac{\partial y_0}{\partial \mu}\right|_{\mu=h=0} = -\int_0^\infty \lambda\sigma[w_0'(\sigma)]^2 \exp[-\lambda w_0(\sigma)]\, d\sigma = -\frac{1}{2\lambda}.$$

This implies that the implicit function theorem is applicable to the first equation in (8.36). Solving it for $\mu = \mu^0(h)$, we get

$$\mu^0(h) = 2\lambda h + O(h^2). \tag{8.38}$$

We substitute this in the second equation in (8.36) and try to find $\tilde{c}_1 = \tilde{c}_1^0(h)$.

It is important to note that under the substitution $\sigma \exp(-\tilde{c}_1/c_0) \to \sigma$ the coefficient $y_1(\mu^0(h), \tilde{c}_1, h)$ in the expansion of $w(\sigma)$ as $\sigma \to \infty$ is transformed into a coefficient $\tilde{y}_1(h)$ independent of $\tilde{c}_1$. Analytically, this means that

$$y_1(\mu^0(h), \tilde{c}_1, h) = \tilde{y}_1(h) \exp(\tilde{c}_1/c_0). \tag{8.39}$$

We use (8.39) in the second equation of (8.36) and observe that $\tilde{y}_1(0) = 1/\lambda$ according to (8.37). Therefore,

$$\tilde{c}_1^0(h) = O(h). \tag{8.40}$$

Returning to the original variables ν and c_1, we get (8.33) from (8.38) and (8.40).

The formulas (8.38) and (8.40) supply solutions of (8.36) that tend to zero as $h \to 0$. We show that there are no other solutions.

Suppose that the first equation in (8.36) has a solution $\mu(h)$, $0 < h \le \overline{h}$, such that

$$\mu(h) \ge m_0\lambda + 1, \tag{8.41}$$

where the positive constant m_0 is taken from the right-hand side of (7.38). In view of (8.39) the second equation in (8.36) is solvable also when (8.41) holds. But then, as follows from the constructions below, equation (7.11) has a solution for which (7.38) is violated, and that is impossible.

For $h = 0$ it follows from (8.34) and (8.35) that

$$\begin{aligned}\frac{1}{\lambda} &= -\int_0^\infty w_0'(\sigma) \exp[-\lambda w_0(\sigma \exp(-\mu))]\, d\sigma \\ &< -\int_0^\infty w_0'(\sigma) \exp[-\lambda w_0(\sigma)]\, d\sigma = \frac{1}{\lambda} \exp[-y_0(\mu, 0)].\end{aligned}$$

Therefore, $y_0(\mu, 0) < 0$ for $0 < \mu \le m_0\lambda + 1$. But this means that $\mu(h) \to 0$ as $h \to 0$. The proof is complete.

It follows from the proof of Lemma 8.2 that the system (8.32) is uniquely solvable not only for $z = z_0$ but also for all z close to z_0. The corresponding solution will be denoted by $(\nu^0(z, h), c_1^0(z, h))$.

LEMMA 8.3. *For* $\nu \le 0$ *equation* (8.24) *does not have a solution with the required properties.*

PROOF. Making the substitutions $\nu \to -\nu$, $u = 1/s$, we transform (8.24) to the form

$$uw_0'' + w_0' = w_0' \exp[-\lambda w_0(uq)], \tag{8.42}$$

where $q = \exp(-\varkappa\nu)$ as before. Further, it follows from the first equality in (8.25) that (8.42) must be supplemented by the initial conditions

$$w_0|_{u\to+0} = 0, \qquad w_0'|_{u\to+0} = 1/\varkappa. \tag{8.43}$$

The local solvability of the problem (8.42), (8.43) is established by passing to the integral equation

$$\omega(u) = \int_0^u \frac{\omega(s)+1/\varkappa}{s}\left\{\exp\left[-\frac{\lambda qs}{\varkappa} - \lambda\int_0^{sq}\omega(\sigma)\,d\sigma\right] - 1\right\} ds,$$

where $\omega = w_0' - 1/\varkappa$. Let $(0, Q)$ be the maximal interval of existence of a solution. We show first that $w_0' > 0$, and then that $Q = \infty$.

The strict increase of w_0 for small u is a consequence of the second equality in (8.43). Assume that there is a u_0 such that $w_0'(u) > 0$ for $0 < u < u_0$, and $w_0'(u_0) = 0$. Then it follows from (8.42) that

$$0 = uw_0'(u)\Big|_0^{u_0} = \int_0^{u_0} w_0'(s)\exp[-\lambda w_0(sq)]\,ds > 0;$$

that is, we get a contradiction.

It follows from the proven monotonicity of w_0 that

$$\lim_{u\to Q-0} w_0(u) = \lim_{u\to Q-0} w_0'(u) = \infty.$$

But then in view of (8.42)

$$J = \int_0^Q w_0'(s)\exp[-\lambda w_0(sq)]\,ds = \infty.$$

On the other hand, since $w_0''(u) < 0$ for $0 < u < Q$, it follows that

$$J \le \int_0^Q w_0'(sq)\exp[-\lambda w_0(sq)]\,ds = 1/\lambda q; \tag{8.44}$$

that is, we come to a contradiction.

Accordingly, the solution of the problem (8.42), (8.43) is defined on the positive semiaxis. Integrating (8.42) over $[0, u]$ and letting u go to ∞, we see that

$$w_0(u) \sim c(\nu)\ln u, \qquad u\to\infty, \tag{8.45}$$

where

$$c(\nu) = \int_0^\infty w_0'(s)\exp[-\lambda w_0(sq)]\,ds. \tag{8.46}$$

We remark that the integral on the right-hand side of (8.46) converges in view of (8.44). The second condition in (8.25) means that

$$c(\nu) = c_0, \tag{8.47}$$

where, recall, $c_0 = \lambda^{-1}\exp(-\lambda h)$. However, it follows from the monotonicity of w_0 that

$$c(\nu) \geq \int_0^\infty w_0'(s) \exp[-\lambda w_0(s)]\, ds = 1/\lambda,$$

and hence (8.47) has no solutions for $\nu \geq 0$. The proof is complete.

8.4. Construction of higher approximations of the solution of the boundary value problem (8.10)–(8.13) on the turning part. Let us return to equation (8.23). The choice of a solution $w(s, \varepsilon)$ of it will be preceded by the following formal procedure. Using the substitutions (8.22) and the representations (8.21) in the series (8.16), and re-expanding with respect to ε, we conclude that the solution of (8.23) of interest to us must have an asymptotic series expansion

$$w = w_0 + \sum_{n=1}^{\infty} \varepsilon^n \sum_{m=0}^{2n} w_{n,m}(s) \ln^m(1/\varepsilon), \tag{8.48}$$

where w_0 is the function in §8.3, defined for all z close to z_0.

This procedure leads to the following algorithm. In (8.23) and (8.11) we set

$$q = q_0 + \sum_{n=1}^{\infty} \varepsilon^n \sum_{m=0}^{2n} q_{n,m} \ln^m(1/\varepsilon), \tag{8.49}$$

$$c_1 = c_1^0 + \sum_{n=1}^{\infty} \varepsilon^n \sum_{m=0}^{2n} c_1^{n,m} \ln^m(1/\varepsilon), \tag{8.50}$$

where $q_0 = \exp(-\varkappa\nu^0(z, h))$, $c_1^0 = c_1^0(z, h)$, and $\nu_0(z, h)$ and $c_1^0(z, h)$ were defined after the proof of Lemma 8.2. We then substitute the series (8.48) in (8.23) in place of w and equate the coefficients of like powers of ε. As a result we get a recursion sequence of differential equations, which we provide with initial conditions starting from the equalities (8.11) and (8.12) with the use of (8.22) and (8.50), for determining the functions $w_{n,m}$. The unknown constants $q_{n,m}$ and $c_1^{n,m}$ are found from the conditions for "matching" the asymptotic formulas (8.16) and (8.48) as $s \to \infty$.

We illustrate the general scheme above with the example $n = 1$, and since all the features of the algorithm are already apparent here, we dwell only on this case.

Proceeding in the way described above, we arrive at the series of initial value problems

$$\begin{aligned} Lw_{1,0} &= \lambda q_{1,0} s w_0'(sq_0) w_0'(s) \exp(-\lambda w_0(sq_0)) \\ &\quad + \left[\exp(-\lambda w_0(sq_0)) - \frac{\lambda}{\varkappa} - \lambda\left(\gamma + \frac{1}{\varkappa}\ln\frac{1}{s}\right)\exp(-\lambda w_0(sq))\right] w_0', \end{aligned} \tag{8.51}$$

$$\left(w_{1,0} - \frac{\lambda c_0}{2\varkappa}\ln^2\frac{1}{s}\right)\Big|_{s\to+0} = c_1^{1,0}, \tag{8.52}$$

$$\left(s w_{1,0}' + \frac{\lambda c_0}{\varkappa}\ln\frac{1}{s}\right)\Big|_{s\to+0} = 0, \tag{8.53}$$

$$(8.54)\quad Lw_{1,1} = \lambda q_{1,1} s w_0'(sq_0) w_0'(s) \exp(-\lambda w_0(sq_0)) - \frac{\lambda}{\varkappa} w_0' \exp(-\lambda w_0(sq_0)),$$

$$(8.55)\quad \left(w_{1,1} - \frac{\lambda c_0}{\varkappa} \ln \frac{1}{s}\right)\bigg|_{s\to+0} = c_1^{1,1},$$

$$(8.56)\quad \left(s w_{1,1}' + \frac{\lambda c_0}{\varkappa}\right)\bigg|_{s\to+0} = 0,$$

$$(8.57)\quad Lw_{1,2} = \lambda q_{1,2} s w_0'(sq_0) w_0'(s) \exp(-\lambda w_0(sq_0)),$$

$$(8.58)\quad w_{1,2}|_{s\to+0} = \frac{\lambda c_0}{2\varkappa} + c_1^{1,2}, \qquad s w_{1,2}'|_{s\to+0} = 0,$$

where

$$Lw = s w'' + w'[1 + \exp(-\lambda w_0(sq_0))] - \lambda w(sq_0) w_0' \exp(-\lambda w_0(sq_0)),$$

whose solvability can be proved by making a transition to certain integral equations (see analogous places in §8.3).

We first dwell on an analysis of the problem (8.51)–(8.53). Let us represent its solution as a sum $w_{1,0}^1 + w_{1,0}^2$, where

$$(8.59)\quad Lw_{1,0}^1 = \left[\exp(-\lambda w_0(sq_0)) - \frac{\lambda}{\varkappa} - \lambda\left(\gamma + \frac{1}{\varkappa}\ln\frac{1}{s}\right)\exp(-\lambda w_0(sq_0))\right] w_0',$$

$$(8.60)\quad \left(w_{1,0}^1 - \frac{\lambda c_0}{2\varkappa} \ln^2 \frac{1}{s}\right)\bigg|_{s\to+0} = 0,$$

$$(8.61)\quad \left(s w_{1,0}' + \frac{\lambda c_0}{\varkappa} \ln \frac{1}{s}\right)\bigg|_{s\to+0} = 0,$$

$$(8.62)\quad Lw_{1,0}^2 = \lambda q_{1,0} s w_0'(sq_0) w_0'(s) \exp(-\lambda w_0(sq_0)),$$

$$(8.63)\quad w_{1,0}^2|_{s\to+0} = c_1^{1,0}, \qquad s w_{1,0}^{2\prime}|_{s\to+0} = 0.$$

Using the asymptotic properties of w_0 as $s \to \infty$ in (8.59)–(8.61) (see Lemma 8.1), we see with the help of the method of undetermined coefficients that $w_{1,0}^1$ has the following representation as $s \to \infty$:

$$(8.64)\quad w_{1,0}^1 = \sum_{k=0}^{2} \ln^k s \sum_{r=1}^{\infty} \frac{\beta_{k,r}^1}{s^r} + \beta_1,$$

where we emphasize that the coefficients $\beta_{k,r}^1$ and β_1 do not depend on the unknown constants $q_{1,0}$ and $c_1^{1,0}$.

For the solution of the problem (8.62) and (8.63) we have the representation

$$(8.65)\quad w_{1,0}^2 = c_1^{1,0} \frac{\partial w_0}{\partial c_1} - \frac{q_{1,0}}{\varkappa q_0} \frac{\partial w_0}{\partial \nu},$$

where $w_0 = w_0(s, \nu, c_1)$ is a solution of the initial value problem (8.24), (8.26), and the derivatives are taken for $c_1 = c_1^0(z, h)$ and $\nu = \nu^0(z, h)$ (the formula (8.65) can be established directly from (8.24) and (8.26) by differentiation with respect to ν and c_1). Therefore, the asymptotic expression for $w_{1,0}^2$ as $s \to \infty$ is easy to obtain: differentiate the asymptotic equality (8.29) with respect to ν and c_1 and take the necessary linear combinations (see (8.65)). In this way we get

$$w_{1,0}^2 = \sum_{k=0}^{1} \ln^k s \sum_{r=1}^{\infty} \frac{\beta_{k,r}^2}{s^r} + \beta_2. \tag{8.66}$$

We now describe a way of finding the unknown constants $q_{1,0}$ and $c_1^{1,0}$. With this goal we use the series (8.49) in (8.16) and (8.21), let $\tau = (1/\varkappa)\ln(1/\varepsilon s)$, and expand again with respect to ε. As a result we see that $w_{n,m}$ has the following asymptotic expression as $s \to \infty$:

$$w_{n,m} = \sum_{k=0}^{2n-m} \ln^k s \sum_{r=1}^{\infty} \frac{\alpha_{n,m,r}^k}{s^r} + \alpha_{n,m}, \qquad n \geq 1, \tag{8.67}$$

where $\alpha_{n,2n} = 0$, and the coefficients $\alpha_{n,m,r}^k \alpha_{n,m}$ depend on the $q_{j,s}$ with indices $j \leq n-1$. We remark also that the coefficients of (8.67) with $n = 1$ and $m = 0$ coincide completely with those of the sum of the series (8.64) and (8.66) if their free terms and the coefficients of $1/s$ coincide. This follows from the fact that both these series, being formal solutions of the same second-order equation, are completely determined by specifying their free terms and the coefficients of $1/s$ (this can be verified by the method of undetermined coefficients). Thus, we get the equations

$$\beta_1 + \beta_2 = \alpha_{1,0}, \qquad \beta_{0,1}^1 + \beta_{0,1}^2 = \alpha_{1,0,1}^0 \tag{8.68}$$

for finding $c_1^{1,0}$ and $q_{1,0}$, where, by (8.65),

$$\beta_2 = -\frac{q_{1,0}}{\varkappa q_0} x_0', \qquad \beta_{0,1}^2 = c_1^{1,0} \frac{\partial x_1}{\partial c_1} - \frac{q_{1,0}}{\varkappa q_0} \frac{\partial x_1}{\partial \nu}, \tag{8.69}$$

x_0 and x_1 are the functions in (8.32), and the derivatives are taken for $\nu = \nu^0(z, h)$ and $c_1 = c_1^0(z, h)$. We recall also that β_1, $\beta_{0,1}^1$, $\alpha_{1,0}$, and $\alpha_{1,0,1}^0$ are independent of the constants $q_{1,0}$ and $c_1^{1,0}$. Therefore, equations (8.68) make up a linear system in $q_{1,0}$ and $c_1^{1,0}$ whose determinant is nonzero by virtue of Lemma 8.2.

The analysis of the initial value problem (8.54)–(8.56) is quite analogous, and we do not dwell on it. A new point arises only in considering the initial value problem (8.57), (8.58), for the solution of which we have a formula analogous to (8.65):

$$w_{1,2} = \left(\frac{\lambda c_0}{2\varkappa} + c_1^{1,2}\right) \frac{\partial w_0}{\partial c_1} - \frac{q_{1,2}}{\varkappa q_0} \frac{\partial w_0}{\partial \nu}. \tag{8.70}$$

From the representation (8.70) for $w_{1,2}$ we get as $s \to \infty$ the asymptotic equality (see the analogous place for $w_{1,0}^2$)

$$w_{1,2} = \sum_{k=0}^{1} \ln^k s \sum_{r=1}^{\infty} \frac{\delta_{k,r}}{s^r} + \delta_0, \tag{8.71}$$

which differs structurally from what is required (see (8.67) for $n = 1$ and $m = 2$). However, after the determination of the constants $q_{1,2}$ and $c_1^{1,2}$ from the equations

$$-\frac{q_{1,2}}{\varkappa q_0} x_0' = 0, \qquad c_1^{1,2} \frac{\partial x_1}{\partial c_1} - \frac{q_{1,2}}{\varkappa q_0} \frac{\partial x_1}{\partial \nu} = \alpha_{1,2,1}^0 \tag{8.72}$$

all the terms of the form $\ln s/s^r$ in (8.71) vanish. Indeed, from the first equation in (8.72) we have that $q_{1,2}=0$. Therefore, by (8.70),

$$w_{1,2}=\left(\frac{\partial c_0}{2\varkappa}+c_1^{1,2}\right)\frac{\partial w_0}{\partial c_1}$$

and the required asymptotic expansion is obtained from (8.29) after differentiation with respect to c_1 and multiplication by $\lambda c_0/\varkappa+c_1^{1,2}$.

The subsequent realization of the algorithm does not lead to difficulties. We note only that in the general case $q_{n,2n}\neq 0$ for $n>1$.

8.5. On verification of the asymptotic formulas. As already mentioned, the problem of verifying the asymptotics of the solution of the problem (8.10)–(8.13) on the turning part $(\lambda_1/\varkappa)\varepsilon\ln(1/\varepsilon)\le t\le\varepsilon^{\lambda_2}$, $0<\lambda_1,\lambda_2<1$, to which there correspond the values

$$(1/\varepsilon)\exp(-\varkappa/\varepsilon^{1-\lambda_2})\le s\le 1/\varepsilon^{1-\lambda_1}, \tag{8.73}$$

involves investigating the properties of the linear equation

$$sh''+h'[1+\exp(-\lambda w_0(sq))]=\lambda h(sq)w_0'\exp(-\lambda w_0(sq)). \tag{8.74}$$

Let $h(s,\tau)$ be a solution of (8.74) with an arbitrary initial condition $\varphi_0(s)\in C(q\tau,\tau)$, $h_0|_{s=\tau}=\delta_0$.

LEMMA 8.4. *The following inequalities hold*:

$$|h(s,\tau)|+\left|\frac{\partial}{\partial s}h(s,\tau)\right|\le M\left(\max_s|\varphi_0(s)|+|\tau\delta_0|\right) \tag{8.75}$$

for $0<s_0\le\tau<s$;

$$|h(s,\tau)|+\left|\frac{\partial}{\partial s}h(s,\tau)\right|\le M\left(\max_s|\varphi_0(s)|+|\tau\delta_0|\right)\ln\frac{1}{\tau} \tag{8.76}$$

for $s\ge s_0$, $0<\tau\le s_0$;

$$|h(s,\tau)|+\left|s\frac{\partial}{\partial s}h(s,\tau)\right|\le M\left(\max_s|\varphi_0(s)|+|\tau\delta_0|\right)\ln\frac{1}{\tau} \tag{8.77}$$

for $0<\tau<s\le s_0$. *Here* $s_0>0$ *is fixed arbitrarily, and* M *is a universal constant.*

For a proof we take the representation

$$\begin{aligned}h(s,\tau)&=\varphi_0(\tau)+\tau\delta_0K_0(s,\tau)\\&\quad+\frac{\lambda}{q}\int_{\tau q}^{\tau}K_0\left(s,\frac{\sigma}{q}\right)w_0'\left(\frac{\sigma}{q}\right)\exp(-\lambda w_0(\sigma))\varphi_0(\sigma)\,d\sigma\\&\quad+\frac{\lambda}{q}\int_{\tau}^{s}K_0\left(s,\frac{\sigma}{q}\right)w_0'\left(\frac{\sigma}{q}\right)\exp(-\lambda w_0(\sigma))h(\sigma,\tau)\,d\sigma,\end{aligned}$$

where

$$K_0(s,\tau)=\frac{1}{\tau}\int_\tau^s\exp\left\{-\int_\tau^\sigma[1+\exp(-\lambda w_0(\delta q))]\,d\delta/\delta\right\}d\sigma,\qquad s\ge\tau,$$
$$K_0(s,\tau)\equiv 0,\qquad s<\tau,$$

and use in it the asymptotic properties of $w_0(s)$ as $s \to 0$ and ∞, which lead to the estimates

$$\begin{aligned} &K_0(s, \tau) \le M, && 0 < s_0 \le \tau < s; \\ &K_0(s, \tau) \le M \ln(1/\tau), && s \ge s_0, \ \tau \le s_0; \\ &K_0(s, \tau) \le M \ln(s/\tau), && 0 < \tau < s \le s_0; \\ &|w_0'(s/q)| \exp(-\lambda w_0(s)) \le M/s^2, && s \ge s_0; \\ &|w_0'(s/q)| \exp(-\lambda w_0(s)) \le M/s^{1-\lambda c_0}, && 0 < s \le s_0. \end{aligned}$$

After this the proof is concluded with the help of the Gronwall-Bellman lemma.

The rest of the arguments are standard: in equation (8.23) we set $w = w_n(s, \varepsilon) + \varepsilon^n \tilde{w}_n + \varepsilon^{n+1} \tilde{w}_{n+1} + h$, $n \ge 1$, $q = q_n(\varepsilon) + \varepsilon^{n+1} r_1$, and in (8.11) we set $c_1 = c_1^n(\varepsilon) + \varepsilon^n r_2$. Here $w_n(s, \varepsilon)$, $q_n(\varepsilon)$, and $c_1^n(\varepsilon)$ are segments of length n of the respective series (8.48), (8.49), and (8.50), $\tilde{w}_n = (\partial w_0/\partial c_1) r_2$, and $\tilde{w}_{n+1} = (\partial w_0/\partial \nu) r_1$, where the functions $\partial w_0/\partial c_1$ and $\partial w_0/\partial \nu$ were introduced in §8.4, and r_1 and r_2 are constants to be determined. Next we supplement the equation obtained for h by the initial conditions

$$h(s, \varepsilon) = \varphi_0(s, \varepsilon), \quad q\tau_0 \le s \le \tau_0, \qquad \dot{h}|_{s=\tau} = \dot{\varphi}_0(\tau_0, \varepsilon),$$

where $\tau_0 = (1/\varepsilon)\exp(-\varkappa/\varepsilon^{1-\lambda_2})$, and

$$\begin{aligned} \varphi_0(s, \varepsilon) = \frac{1}{\lambda\varepsilon}\left[\exp\lambda\left(z + \frac{\varepsilon}{\varkappa}\ln\frac{1}{s\varepsilon}\right) - \exp\lambda z\right] - c_0 \ln\frac{1}{\varepsilon} \\ + c_1 - \varepsilon^n \tilde{w}_n - \varepsilon^{n+1}\tilde{w}_{n+1} - w_n, \end{aligned}$$

and we pass from a differential equation to an integral equation in the usual way. By Lemma 8.4, the latter gives rise to an integral operator that transforms a suitable set of functions into itself and is a contraction. In this way we get a function $h(s, r_1, r_2, \varepsilon)$ which admits the estimate

$$|h| + \left|\alpha(s)\frac{\partial}{\partial s}h\right| \le M\varepsilon^{(n+3)\lambda_2 - 2} \tag{8.78}$$

for all s in the interval (8.73). Here $\alpha(s) = s$, $0 \le s \le 1$, and $\alpha(s) \equiv 1$ for $s \ge 1$. Moreover, estimates analogous to (8.78) are valid also for the derivatives of the functions h and $\partial h/\partial s$ with respect to r_1 and r_2.

We do not give the details of the verification of the asymptotic formulas for the solution of the problem (8.10)–(8.13) on the rapid motion part (that is, for $0 \le \tau \le (\lambda_1/\varkappa)\ln(1/\varepsilon)$), because here the matter reduces to an analysis of the simple equation $\ddot{v}_0 = \varkappa \dot{v}_0$. We mention only that the constants r_1 and r_2 introduced earlier are determined from (8.15).

In conclusion we underscore once more that in view of (8.78), to construct the solution of the basic boundary value problem (8.10)–(8.13) with specified accuracy it is necessary to know its explicit formal asymptotics.

8.6. Computation of the parameters $\gamma = \gamma(z, \varepsilon)$ and $z = z(\varepsilon)$. It follows from equation (7.11) that for $h - \varepsilon^\lambda \le t \le h + \varepsilon\nu$, $0 < \lambda < 1$, the relaxation cycle $N(t, \varepsilon)$ is given with exponential accuracy by the formula

$$N(t, \varepsilon) = \exp\lambda(t + z - \psi(t - h, \varepsilon) + \psi(-\varepsilon^\lambda, \varepsilon)), \tag{8.79}$$

where $\psi(t, \varepsilon)$ is the solution of the basic boundary value problem (8.10)–(8.13). Setting $t = h + \varepsilon\nu$ in (8.79) and using the asymptotic representation (8.16), we get

$$N(h + \varepsilon\nu, \varepsilon) = \exp\lambda\left(h + \varepsilon\nu + z + \varepsilon\sum_{k=0}^{\infty}\varepsilon^k r_k\right), \tag{8.80}$$

where the constants r_k are the same as in (8.21). Comparing (8.80) and (8.7), we get an equality for determining γ:

$$\gamma = \nu + \sum_{k=0}^{\infty}\varepsilon^k r_k, \tag{8.81}$$

and using $\nu = (-1/\varkappa)\ln q$ in it, where q is the series (8.49), we finally deduce that

$$\gamma + \gamma_0(z) = \sum_{n=1}^{\infty}\varepsilon^n\sum_{m=0}^{2n}\gamma_{n,m}(z)\ln^m\frac{1}{\varepsilon}, \tag{8.82}$$

where $\gamma_0(z) = \nu^0(z, h) - (1/\varkappa)\exp[\varkappa\nu^0(z, h)]$. We note further that the equality $q_{1,2} = 0$ (see §8.4) implies that $\gamma_{1,2}(z) \equiv 0$.

Using the algorithms presented above, we easily get the complete asymptotics of $N(t, \varepsilon)$ on the whole spike part, since its computation reduces to taking certain integrals of formulas already obtained (see §8.7, where the principal asymptotic expression is given). Therefore, we do not write out the corresponding series. We confine ourselves to the formula for the remaining constant z:

$$z = z_0 + z_1\varepsilon\ln\frac{1}{\varepsilon} + \sum_{n=1}^{\infty}\varepsilon^n\sum_{m=0}^{2n-2}z_{n,m}\ln^m\frac{1}{\varepsilon}, \tag{8.83}$$

where $z_1 = c_0/\lambda(1+h)$, which can be determined from the equation

$$\lambda(1 + h + \varepsilon\gamma) - \lambda\int_{-1+h+\varepsilon\nu}^{h+\varepsilon\nu}N(s, \varepsilon)\,ds + \frac{\lambda}{\varepsilon}\int_{-1+h+\varepsilon\nu}^{\varepsilon\nu}N(s, \varepsilon)\,ds = 0.$$

8.7. Asymptotic formulas for the relaxation cycle on the spike part. In this subsection we give asymptotic formulas needed for studying the stability of a relaxation cycle.

It follows from §§8.1 and 8.2 that on the rapid motion part

$$\begin{gathered} N(t, \varepsilon) = \varepsilon\exp\varkappa\tau + O(\varepsilon^{2-2\lambda_1}), \qquad 0 < \lambda_1 < 1, \\ t = \varepsilon\tau, \qquad 0 \le \tau \le \frac{\lambda_1}{\varkappa}\ln\frac{1}{\varepsilon}, \end{gathered} \tag{8.84}$$

where $\varkappa = \lambda\exp\lambda(z_0 + h)$ here and below. It follows from §§8.3 and 8.4 that on the turning part

$$\begin{gathered} N(t, \varepsilon) = -\varkappa s w_0'(s) + O(\varepsilon^{\lambda_0}), \qquad 0 < \lambda_0 < 1, \\ t = \frac{\varepsilon}{\varkappa}\ln\frac{1}{\varepsilon s}, \qquad \frac{1}{\varepsilon}\exp(-\varkappa\varepsilon^{\lambda_0 - 1}) \le s \le \varepsilon^{1-\lambda_1}; \end{gathered} \tag{8.85}$$

here and below, w_0 is taken for $z = z_0$. Finally, on the slow motion part (that is, for $\varepsilon^{\lambda_0} \le t \le h + \varepsilon\nu$), the results of the preceding section imply that

$$N(t, \varepsilon) = \{1 + O(\varepsilon\ln(1/\varepsilon))\}\exp\lambda(z_0 + t). \tag{8.86}$$

Discarding the terms of order $\exp(-m/\varepsilon)$, $m > 0$, from (7.11), we conclude that for $h + \varepsilon\nu \le t \le 2h + \varepsilon\nu$

$$N(t, \varepsilon) = N(h + \varepsilon\nu, \varepsilon) \exp\left\{\lambda(t - h - \varepsilon\nu) - \frac{\lambda}{\varepsilon}\int_{\varepsilon\nu}^{t-h} N(s, \varepsilon)\, ds\right\}. \tag{8.87}$$

We set $t - h - \varepsilon\nu = \varepsilon\tau$ in (8.87) and use (8.86) (with $t = h + \varepsilon\nu$) and (8.84). As a result, for the beginning of the drop part we get the asymptotic formula

(8.88)

$$N(t, \varepsilon) = \left\{1 + \varepsilon\lambda\left[\tau - \frac{1}{\varkappa}\exp\varkappa(\tau + \nu) + \frac{1}{\varkappa}\exp\varkappa\nu\right] + O(\varepsilon^{2-2\lambda_1})\right\}\exp\lambda(z_0 + h),$$
$$0 \le \tau \le \frac{\lambda_1}{\varkappa}\ln\frac{1}{\varepsilon}.$$

To get the asymptotic formula at the end of the drop part we make the substitution (8.22) in (8.87) and use the equality

$$w(s, \varepsilon) = w_0(s) + O(\varepsilon^{2\lambda_0 - 1}), \qquad \frac{1}{\varepsilon}\exp(-\varkappa\varepsilon^{\lambda_0 - 1}) \le s \le \frac{1}{\varepsilon^{1-\lambda_1}}.$$

Finally, we obtain

$$\begin{gathered} N(t, \varepsilon) = \{1 + O(\varepsilon^{2\lambda_0 - 1})\}\exp\lambda[z_0 + h - w_0(sq)], \\ t - h - \varepsilon\nu = \frac{\varepsilon}{\varkappa}\ln\frac{1}{s\varepsilon}, \qquad \frac{1}{\varepsilon}\exp(-\varkappa\varepsilon^{\lambda_0 - 1}) \le s \le \frac{1}{\varepsilon^{1-\lambda_1}}, \end{gathered} \tag{8.89}$$

where, as above, $q = \exp(-\varkappa\nu)$.

Let us sum up the foregoing in a theorem.

THEOREM 8.1. *For sufficiently small h the asymptotic equalities* (8.84)–(8.86) *and* (8.88)–(8.89) *are valid on the spike part of the relaxation cycle of* (7.11) *generated by the exponentially stable equilibrium state* (7.14) *of the mapping* (7.12), *and its period satisfies the asymptotic equality*

$$T = 1 + h + \varepsilon h\delta(\varepsilon, h),$$

where $\delta_0(h) = \delta(0, h) = 2h + O(h^2)$, $\delta(\varepsilon, 0) \to 0$ as $\varepsilon \to 0$, and

$$\delta(\varepsilon, h) = \delta_0(h) + \sum_{n=1}^{\infty}\varepsilon^n\sum_{m=0}^{2n}\delta_{n,m}(h)\ln^m\frac{1}{\varepsilon}, \tag{8.91}$$

with $\delta_{1,2}(h) \equiv 0$.

We remark that (8.91) is obtained by substituting (8.83) in the expression $\nu = (-1/\varkappa)\ln q$, where q is the series (8.49), and expanding again with respect to ε. The equality $\delta_{1,2} \equiv 0$ is a consequence of the fact that $q_{1,2} = 0$ in (8.49).

It remains an open question as to whether Theorem 8.1 is valid for arbitrary $h < 1$.

Another open question has to do with constructing the asymptotics of the relaxation cycle corresponding to an exponentially stable cycle (7.13) of (7.12) for $k \ge 1$. The general course of action is clear: for a given spike of the cycle determine its drop, then write a formula for the next spike, and so on, and finally require that the next spike coincide with the first one. This program was essentially realized in §8.1, and this enabled us to obtain the basic boundary value problem (8.10)–(8.13). Its realization in a more complicated case leads to the following conclusions. The single equation (8.9) is replaced by a system of $k+1$ equations that allows us to construct the asymptotics of the cycle on the rapid motion parts. However, we cannot manage with the single substitution (8.22) in studying the turning parts. We have to resort to the method of several scales, and hence we study a system of partial differential equations instead of the single equation (8.24). For example, for $k = 1$ we have instead of equation (8.9) the system

$$\tag{8.92} \begin{aligned} \dot{\varphi}_1 &= \lambda\left[1 + \frac{a}{\varepsilon}\exp\left\{\lambda(t + z_2 + h + \varepsilon\gamma_2) - \lambda\int_{\varepsilon\nu_2}^{t+\varepsilon\nu_2} \varphi_2(s, \varepsilon)\, ds\right\}\right]\varphi_1, \\ \dot{\varphi}_2 &= \lambda\left[1 + \frac{a}{\varepsilon}\exp\left\{\lambda(t + z_1 + h + \varepsilon\gamma_1) - \lambda\int_{\varepsilon\nu_1}^{t+\varepsilon\nu_1} \varphi_1(s, \varepsilon)\, ds\right\}\right]\varphi_2, \end{aligned}$$

where $\varphi_1(0, \varepsilon) \equiv \varphi_2(0, \varepsilon) \equiv 1$, and the parameters z_1, z_2, γ_1, γ_2, ν_1, and ν_2 have the same meaning as z, γ, and ν in (8.9). Next, setting

$$v_j = \int_{\varepsilon\nu_j}^{t} \varphi_j(s, \varepsilon)\, ds, \qquad j = 1, 2,$$

and then $u_j = \varepsilon\exp(\varkappa_j t/\varepsilon)$ and $\varkappa_j = \lambda\exp\lambda(z_j + h)$, $j = 1, 2$, in (8.92), we get the following system, which is analogous to (8.24):

$$\begin{aligned} Lw_1 &= \varkappa_2\left(\varkappa_1 u_1\frac{\partial w_1}{\partial u_1} + \varkappa_2 u_2\frac{\partial w_1}{\partial u_2}\right)\exp[-\lambda w_2(u_1 q_{21}, u_2 q_{22})], \\ Lw_2 &= \varkappa_1\left(\varkappa_1 u_1\frac{\partial w_2}{\partial u_1} + \varkappa_2 u_2\frac{\partial w_2}{\partial u_2}\right)\exp[-\lambda w_1(u_1 q_{11}, u_2 q_{12})], \end{aligned}$$

where $q_{ij} = \exp(\varkappa_i \nu_j)$, $i, j = 1, 2$, and

$$L = \varkappa_1^2 u_1^2\frac{\partial^2}{\partial u_1^2} + 2\varkappa_1\varkappa_2 u_1 u_2\frac{\partial^2}{\partial u_1\partial u_2} + \varkappa_2^2 u_2^2\frac{\partial^2}{\partial u_2^2} + \varkappa_1^2 u_1\frac{\partial}{\partial u_1} + \varkappa_2^2 u_2\frac{\partial}{\partial u_2}.$$

9. An investigation of stability for the simplest relaxation cycle

9.1. A lemma on finiteness of the spectral radius of the monodromy operator. We proceed to an analysis of the variational equation

$$\tag{9.1} \dot{v} = \lambda\left[\frac{a}{\varepsilon}v(t-1) - \frac{1}{\varepsilon}v(t-h)\right]N(t, \varepsilon) + \frac{\dot{N}(t, \varepsilon)}{N(t, \varepsilon)}v,$$

where $N(t, \varepsilon)$ is one of the periodic solutions of equation (7.11) referred to in Theorem 8.1. As its phase space we take $E = L(-1, 0) \times R$; that is, for $-1 \le t \le 0$ the initial function belongs to $L(-1, 0)$ and the value of the solution at $t = 0$ is given. Denote by V the monodromy operator of equation (9.1).

LEMMA 9.1. *There exist an $\varepsilon_0 > 0$ and a positive constant M_0 such that for* $0 < \varepsilon < \varepsilon_0$

$$\tag{9.2} \|V^n\|_{E\to E} \le M_0^n \ln(1/\varepsilon), \qquad n = 1, 2, \dots.$$

PROOF. For $0 \le t \le T$, where $T = 1 + h + \varepsilon\nu$ is the period of $N(t, \varepsilon)$, we denote by $v_0(t, \varepsilon)$ the solution of (9.1) with initial function $v_0(t)$ normalized by 1 in E.

By (9.1),

$$v_0(t, \varepsilon) = v_0(0)\frac{N(t, \varepsilon)}{\varepsilon} + \lambda\frac{N(t, \varepsilon)}{\varepsilon}\int_0^t [av_0(s-1) - v_0(s-h)]\, ds, \quad 0 \le t \le h, \tag{9.3}$$

$$v(t, \varepsilon) = v_0(h, \varepsilon)\frac{N(t, \varepsilon)}{N(h, \varepsilon)} + \lambda\frac{N(t, \varepsilon)}{\varepsilon}\int_h^t [av_0(s-1, \varepsilon) - v_0(s-h, \varepsilon)]\, ds, \quad h \le t \le T. \tag{9.4}$$

It follows from (9.3) and (9.4) that for $h+\delta \le t \le T-\delta$, where $\delta > 0$ is small but fixed, $v_0(t, \varepsilon)$ has order $\exp(-m/\varepsilon)$, $m > 0$, uniformly with respect to the initial conditions. Therefore, in order to estimate $v_0(t, \varepsilon)$ in the metric of $L(T-1, T)$ it suffices to estimate the integrals

$$J_1 = \int_{h+\varepsilon\nu}^{h+\delta} |v_0(t, \varepsilon)|\, dt, \qquad J_2 = \int_{T-\delta}^{T} |v_0(t, \varepsilon)|\, dt. \tag{9.5}$$

Using (9.3) in (9.4), we get

$$J_1 \le \frac{M_1}{\varepsilon}\int_{h+\varepsilon\nu}^{h+\delta} N(t, \varepsilon)\, dt + \frac{M_2}{\varepsilon^2}\int_{h+\varepsilon\nu}^{h+\delta} N(t, \varepsilon)\int_h^t N(s-h, \varepsilon)\, ds\, dt. \tag{9.6}$$

Here and below, $M_1, M_2, \ldots$ are universal positive constants. It follows from the asymptotic properties of $N(t, \varepsilon)$ that the first term on the right-hand side of (9.6) has order $\ln(1/\varepsilon)$, and the second is bounded as $\varepsilon \to 0$. Therefore,

$$J_1 \le M_3 \ln(1/\varepsilon). \tag{9.7}$$

Proceeding to an estimate of the second integral in (9.5), we note the inequality

$$|v_0(t, \varepsilon)| \le \frac{M_4}{\varepsilon}\ln\frac{1}{\varepsilon}N(t, \varepsilon), \qquad 0 \le t \le 1+\varepsilon\nu, \tag{9.8}$$

which follows from (9.3) and (9.4). Using the equality

$$\begin{aligned} &\int_h^t [av_0(s-1, \varepsilon) - v_0(s-h, \varepsilon)]\, ds \\ &\quad = a\int_{-1+h}^{0} v_0(s)\, ds - \varepsilon\int_0^{t-1} v_0(s, \varepsilon)\, ds - \int_{t-1}^{t-h} v_0(s, \varepsilon)\, ds \end{aligned} \tag{9.9}$$

in (9.4) along with the inequality (9.8), we get

$$J_2 \le M_5\varepsilon\ln\frac{1}{\varepsilon} + \frac{M_6}{\varepsilon^2}\ln\frac{1}{\varepsilon}\int_{-\delta}^{0} N(t, \varepsilon)\int_{t-1}^{t-h} N(s, \varepsilon)\, ds\, dt \le M_7\varepsilon\ln^2\frac{1}{\varepsilon}, \tag{9.10}$$

where the asymptotic properties of $N(t, \varepsilon)$ are used again.

To estimate $v_0(T, \varepsilon)$, we first note the inequalities

$$|v_0(h, \varepsilon)| \le M_8/\varepsilon, \qquad \int_0^{h+\varepsilon\nu} |v_0(s, \varepsilon)|\, ds \le M_9/\varepsilon, \tag{9.11}$$

which follow from (9.3) and (9.4). It then remains to use (9.9), (9.7), and (9.11) in (9.4), obtaining

$$|v_0(T, \varepsilon)| \le M_{10}\ln(1/\varepsilon). \tag{9.12}$$

Thus, by (9.7), (9.10), and (9.12),

$$\|V\|_{E\to E} \le M_0 \ln(1/\varepsilon)\,; \tag{9.13}$$

that is, the assertion of the lemma holds for $n = 1$.

Denote by $v_1(t, \varepsilon)$ the solution of equation (9.1) with initial function $v_0(T+t, \varepsilon)$, $-1 \le t \le 0$. It follows from the formula for $v_1(t, \varepsilon)$ analogous to (9.3) and from the inequality (9.13) that for $0 \le t \le h$

$$|v_1(t, \varepsilon)| \le \frac{N(t, \varepsilon)}{\varepsilon}\left[\gamma(\varepsilon) + \lambda \int_{t+h+\varepsilon\nu}^{t+1+\varepsilon\nu} |\nu_0(s, \varepsilon)|\, ds\right], \tag{9.14}$$

where $\gamma(\varepsilon)$ is bounded as $\varepsilon \to 0$. From (9.14) and (9.10) we derive the estimate

$$|v_1(h, \varepsilon)| \le M_{11}/\varepsilon, \tag{9.15}$$

which is analogous to the first inequality in (9.11). If we then write out for $v_1(t, \varepsilon)$ the formula analogous to (9.4) and use (9.14) and (9.15), we get

$$\begin{aligned}
&\int_{h+\varepsilon\nu}^{h+\delta} |v_1(t, \varepsilon)|\, dt \\
&\quad\le \frac{M_{12}}{\varepsilon}\int_{h+\varepsilon\nu}^{h+\delta} N(t, \varepsilon)\, dt + \frac{M_{13}}{\varepsilon^2}\ln\frac{1}{\varepsilon}\int_{h+\varepsilon\nu}^{h+\delta} N(t, \varepsilon)\int_0^{t-h} N(s, \varepsilon)\, ds\, dt \\
&\quad\le M_{14}\ln\frac{1}{\varepsilon},
\end{aligned} \tag{9.16}$$

where the asymptotic properties of $N(t, \varepsilon)$ were used again. We conclude from (9.16) that the quantity for $v_1(t, \varepsilon)$ analogous to J_1 has been estimated. The quantity analogous to J_2 and $|v_1(t, \varepsilon)|$ are estimated precisely as in the case of $v_0(t, \varepsilon)$. It is essential that in this way we arrive at the inequality

$$\|V^2\|_{E\to E} \le M_{15}M_0\ln\frac{1}{\varepsilon}, \tag{9.17}$$

in which M_{15} does not depend on the M_0 in (9.13). Therefore, in the consideration of V^3 we get an estimate of the form (9.17) with M_{15} replaced by M_{15}^2, and so on. Equating M_{15} and M_0, we arrive at (9.2). The proof is complete.

The following consequence of Lemma 9.1 will be of interest below.

LEMMA 9.2. *The multipliers of equation* (9.1) *lie in some disk whose radius is independent of* ε.

9.2. Computation of the limit of the monodromy operator on smooth initial functions. In (9.1) we make the substitution $v/N \to \dot v$ and pass to the associated equation

$$\dot v = -\frac{\lambda}{\varepsilon}N(t, \varepsilon)[av(t+1) - v(t+h)], \tag{9.18}$$

with $C(0, 1)$ as its phase space. Let W be the monodromy operator of equation (9.18). We find its limit as $\varepsilon \to 0$, assuming that the initial functions $v_0(t)$ are in $C_0(0, 1)$, where $C_0(0, 1)$ is the linear subspace of functions in $C(0, 1)$ having finite derivatives at $t = 0$.

LEMMA 9.3. *For* $v_0(t) \in C_0(0, 1)$

$$\lim_{\varepsilon \to 0} W[v_0(t)] = v_0(0)\varphi(t), \tag{9.19}$$

where

$$\varphi(t) = \begin{cases} 1 - \exp \lambda(z_0 + h) + \exp \lambda(z_0 + t), & 0 \le t \le h, \\ 1, & h \le t \le 1. \end{cases}$$

PROOF. For a fixed function $v_0(t)$ we begin to integrate equation (9.18) by steps. In particular, for $-h \le t \le 0$ we arrive at the formula

$$v(t, \varepsilon) = v_0(0) + \frac{\lambda}{\varepsilon} \int_t^0 N(s, \varepsilon)[av_0(s+1) - v_0(s+h)]\, ds. \tag{9.20}$$

It follows from (9.20) and the asymptotic properties of $N(t, \varepsilon)$ that, uniformly with respect to $-h \le t \le -\gamma_0 \varepsilon \ln(1/\varepsilon)$,

$$v(t, \varepsilon) = v_0(0) + \varepsilon \frac{\lambda}{\varkappa}[v_0(1) - v_0(h)] + o(\varepsilon), \tag{9.21}$$

where $\varkappa$ is from (8.14) with $z = z_0$, and $\gamma_0 > 0$ is fixed arbitrarily; moreover, uniformly with respect to $-\gamma_0 \varepsilon \ln(1/\varepsilon) \le t \le 0$,

$$v(t, \varepsilon) = v_0(0) + O(\varepsilon). \tag{9.22}$$

Integrating equation (9.18) further by steps and again using the properties of $N(t, \varepsilon)$, we see that the asymptotic equality (9.21) is preserved for $-1 + \gamma_1 \varepsilon \ln(1/\varepsilon) \le t \le -h$, where $\gamma_1 > 0$ is arbitrarily large but fixed.

For $-1 \le t \le -1 + \gamma_1 \varepsilon \ln(1/\varepsilon)$ we have

$$v(t, \varepsilon) = v\left(-1 + \gamma_1 \varepsilon \ln \frac{1}{\varepsilon}, \varepsilon\right) + \frac{\lambda}{\varepsilon} \int_t^{-1+\gamma_1 \varepsilon \ln(1/\varepsilon)} N(s, \varepsilon)[av_0(s+1) - v(s+h, \varepsilon)]\, ds.$$

For the s of interest to us,

$$|v_0(s+1) - v_0(0)| \le M(1+s).$$

From this and (9.21) we deduce that, uniformly with respect to the t under consideration,

$$v(t, \varepsilon) = v_0(0) + O\left(\varepsilon \ln^2 \frac{1}{\varepsilon}\right). \tag{9.23}$$

For $-1 - h \le t \le -1$ we see by using the equalities (9.21)–(9.23) in the formula

$$v(t, \varepsilon) = v(-1, \varepsilon) + \frac{\lambda}{\varepsilon} \int_t^{-1} N(s, \varepsilon)[av(s+1, \varepsilon) - v(s+h, \varepsilon)]\, ds$$

that, uniformly with respect to the t under consideration,

$$v(t, \varepsilon) = [1 - \exp \lambda(z_0 + h) + \exp \lambda(z_0 + t + 1 + h)]v_0(0) + o(1). \tag{9.24}$$

It remains to write the formula analogous to (9.20) for the initial moment of time $-1 - h$ and to see that the equality (9.24) holds up to the moment of time $-1 - h - \varepsilon\nu$ on the basis of what has already been obtained. The lemma is proved.

9.3. Nonsmooth initial conditions. An analysis of the proof of Lemma 9.3 confirms that only the values of $v_0(t)$ for $0 \le t \le A\varepsilon \ln(1/\varepsilon)$ have an essential influence on the character of behavior of $v(t, \varepsilon)$ for $-T \le t \le 0$, where the positive parameter A is sufficiently large. Therefore, we proceed to a study of the solutions $v(t, \varepsilon)$ of equation (9.18) with initial functions $v_0(t, \varepsilon)$ that change essentially for the indicated t, with $v_0(0, \varepsilon) = 0$ and $v_0(t, \varepsilon) \equiv 1$ for $A\varepsilon \ln(1/\varepsilon) \le t \le 1$.

Applying the step method to equation (9.18), we conclude that each solution with the indicated initial condition is close to zero for $-1 + A\varepsilon \ln(1/\varepsilon) \le t \le 0$, while for $-1 \le t \le -1 + A\varepsilon \ln(1/\varepsilon)$ we have up to asymptotically small terms that

$$v(t, \varepsilon) = \frac{\lambda a}{\varepsilon} \int_t^{-1+A\varepsilon \ln(1/\varepsilon)} N(s, \varepsilon) v_0(s+1, \varepsilon)\, ds.$$

For $-T + A\varepsilon \ln(1/\varepsilon) \le t \le -1$ the solution $v(t, \varepsilon)$ is close to $\mu = \nu(-1, \varepsilon)$, while for $-T \le t \le -T + A\varepsilon \ln(1/\varepsilon)$ we have, up to asymptotically small terms,

$$v(t, \varepsilon) = \mu - \frac{\lambda^2 a}{\varepsilon^2} \int_t^{-T+A\varepsilon \ln(1/\varepsilon)} N(s, \varepsilon) \int_{s+h}^{-1+A\varepsilon \ln(1/\varepsilon)} N(\sigma, \varepsilon) v_0(\sigma+1, \varepsilon)\, d\sigma\, ds.$$

We are interested in the eigenfunctions of the monodromy operator W. Therefore, we set $t = -T + \tau$, $0 \le \tau \le A\varepsilon \ln(1/\varepsilon)$, in the last expression and require that the result be equal to $\mu v_0(\tau, \varepsilon)$. After differentiating with respect to τ, we get

$$\mu \dot{v}_0(\tau, \varepsilon) = \frac{\lambda^2 a}{\varepsilon^2} N(\tau, \varepsilon) \int_{\tau - \varepsilon\nu}^{A\varepsilon \ln(1/\varepsilon)} N(s-1, \varepsilon) v_0(s, \varepsilon)\, ds.$$

Differentiating once more with respect to τ, we arrive at the differential equation

$$\mu\left(\ddot{v}_0 - \frac{\dot{N}}{N}\dot{v}_0\right) + \frac{\lambda^2 a}{\varepsilon^2} N(\tau, \varepsilon) N(\tau+h, \varepsilon) v_0(\tau - \varepsilon\nu, \varepsilon) = 0. \tag{9.25}$$

The formulas (8.85), (8.89) show that the coefficients of (9.25) change essentially for

$$\tau = \frac{\varepsilon}{\varkappa} \ln \frac{1}{\varepsilon s}. \tag{9.26}$$

Making the substitution (9.26) in (9.25) and letting ε go to zero, we get the limit equation

$$\mu[s v_0'' + (1 + \exp(-\lambda w_0(sq))) v_0'] - \lambda v_0(s/q) w_0'(s) \exp(-\lambda w_0(s)) = 0. \tag{9.27}$$

Here it is convenient to make the substitution $s = 1/u$, which enables us to write the limit equation in the form

$$\mu[u v_0'' + (1 - \exp(-\lambda w_0(u/q))) v_0'] + \lambda v_0(uq) \omega_0'(u) \exp(-\lambda \omega_0(u)) = 0, \tag{9.28}$$

where $\omega_0(u) = w_0(1/u)$. It follows from the method of deriving (9.28) that

$$v_0|_{u \to +0} = 0, \qquad v_0|_{u \to \infty} = 1. \tag{9.29}$$

Thus, the limit values of the multipliers of equation (9.18) with the initial functions under consideration are found from the boundary value problem (9.28), (9.29).

We set

$$G(u, \sigma) = \begin{cases} [K(\sigma)/K'(\sigma)](\lambda/\sigma)\omega_0'(\sigma) \exp(-\lambda\omega_0(\sigma)), & 0 \le \sigma \le u, \\ [K(u)/K'(\sigma)](\lambda/\sigma)\omega_0'(\sigma) \exp(-\lambda\omega_0(\sigma)), & u \le \sigma, \end{cases}$$

where

$$K(u) = \int_0^u \exp\left\{\frac{1}{s}\int_0^s [\exp(-\lambda\omega_0(\tau/q)) - 1]\,d\tau\right\} ds,$$

and pass from the boundary value problem (9.28), (9.29) to the integral equation

$$\mu v_0(u) = \int_0^\infty G(u, \sigma) v_0(\sigma q)\,d\sigma. \tag{9.30}$$

The right-hand side of (9.30) gives rise to a compact and positive operator L on $C(0, \infty)$ (nontrivial nonnegative functions are transformed into strictly positive functions for $u > 0$). Moreover, it follows from the properties of solutions of second-order linear differential equations that the operator L is oscillatory (in the symmetric case such integral operators were studied by Kellogg [77], [78], whose results were extended in [10] to the nonsymmetric case; the general concept of an oscillatory operator is contained in the note [71]). This implies that its spectrum consists of countably many positive and simple eigenvalues

$$\mu_1 > \mu_2 > \cdots > \mu_n > \cdots, \tag{9.31}$$

where an eigenfunction having exactly $n-1$ zeros on $(0, \infty)$ corresponds to the eigenvalue μ_n.

LEMMA 9.4. $\mu_1 = 1$.

PROOF. By passing to the corresponding integral equation we verify that when $\mu = 1$ the Cauchy problem

$$v_0|_{u\to+0} = 0, \qquad v_0'|_{u\to+0} = 1 \tag{9.32}$$

is solvable for equation (9.28). We show that the solution $v_0(u)$ of this problem is positive for $u > 0$ and that it has a finite positive limit as $u \to \infty$.

For $\mu = 1$ we deduce from (9.23) and (9.28) that

$$uv_0'(u) = v_0(uq)\exp(-\lambda\omega_0(u)) + \int_{uq}^u v_0'(\sigma)\exp(-\lambda\omega_0(\sigma/q))\,d\sigma,$$

from which it follows without difficulty that $v_0'(u) > 0$ for $u > 0$. This leads us to conclude that

$$uv_0'(u) \le v_0(u)\exp(-\lambda\omega_0(u)). \tag{9.33}$$

From (9.33) and the asymptotic properties of $\omega_0(u)$ it follows that the function $v_0(u)$ is bounded as $u \to \infty$. The lemma is proved.

9.4. Limit values of the multipliers. It follows from the results in the preceding two subsections that the numbers zero and $1 - \exp\lambda(z_0 + h) + \exp\lambda z_0 = \mu_0$ and the chain of numbers (9.31) are limit values for the multipliers of equation (9.18).

LEMMA 9.5. *The indicated numbers exhaust all the limit values of the multipliers of equation* (9.18) *as* $\varepsilon \to 0$.

PROOF. According to Lemma 9.2, each multiplier converges to a finite limit at least on some sequence $\varepsilon_k \to 0$. Suppose that this limit μ is different from the numbers indicated above.

By Lemma 9.2, each Lyapunov-Floquet solution is bounded in the C-metric as $\varepsilon \to 0$. From this it automatically follows that their initial conditions are representable as a sum of a smooth and a nonsmooth component. Integrating (9.18) separately with each of them and then adding the results, we get solvability of the boundary value problem

$$\begin{aligned}\mu(sx''+x')+\left[\mu x'+\frac{\mu-\mu_0}{\varkappa s}c\dot{\varphi}(0)\right]\exp(-\lambda w_0(sq))\\ -\lambda x\left(\frac{s}{q}\right)w_0'(s)\exp(-\lambda w_0(s))=\frac{\mu-\mu_0}{\varkappa^2 s}c\ddot{\varphi}(0),\end{aligned} \tag{9.34}$$

$$x|_{s\to+0}=1,\qquad x|_{s\to\infty}=0, \tag{9.35}$$

where φ is the function in Lemma 9.3, and the parameter c is connected with the presence of the smooth component in the initial condition. An uncomplicated analysis of equation (9.34) shows that each solution of it grows unboundedly as $s\to+0$ according to the law

$$\frac{\mu-\mu_0}{\varkappa^2\mu}c\ddot{\varphi}(0)\ln^2 s.$$

Hence, the boundary value problem (9.34), (9.35) is solvable only for $\mu=\mu_0$ or $c=0$, which is a contradiction. The lemma is proved.

We consider the T-periodic solution $v_0(t,\varepsilon)$ of equation (9.18) corresponding to the unit multiplier. The technique worked out in §9.3 enables us to compute the limit

$$-\lambda\varkappa\int_0^\infty v_0(uq)w_0'(u)\exp(-\lambda\omega_0(u))\,du \tag{9.36}$$

of the Hale inner product

$$\dot{N}(0,\varepsilon)v_0(0,\varepsilon)+\lambda a\int_{-1}^0\dot{N}(s,\varepsilon)v_0(s+1,\varepsilon)\,ds-\lambda\int_{-h}^0\dot{N}(s,\varepsilon)v_0(s+h,\varepsilon)\,ds$$

of the functions $v_0(t,\varepsilon)$ and $\varepsilon\dot{N}(t,\varepsilon)/N(t,\varepsilon)$ as $\varepsilon\to 0$, where $v_0(u)$ is the function in Lemma 9.4. The number (9.36) is nonzero. Therefore, the unit multiplier of equation (9.18) is simple, and in view of §9.3 none of the other multipliers tend to 1 as $\varepsilon\to 0$.

We recall that on the spike part all the cycles under consideration have the same principal asymptotics. The following statement therefore sums up the foregoing facts.

THEOREM 9.1. *Under the conditions of Theorem* 8.1 *the set* S_0 *in Theorem* 7.1 *contains an initial condition of a cycle that is orbitally exponentially stable.*

We can now supplement §7.4, where, recall, we discussed the question of the limits of applicability of Theorem 7.1. It turns out that the domain of attraction of even the simplest relaxation cycle (referred to in Theorem 9.1) of equation (7.11) shrinks sharply as ε decreases. However, this circumstance does not diminish the value of Theorem 7.1 and the subsequent results, since with their help we get a good idea of the dynamics of equation (7.11) for moderate values of ε. We point out that all this is not typical; that is, a different situation is usually observed in the asymptotic theory of oscillations (see, for example, §5.4).

9.5. On the dynamics of insect populations with the migration factor taken into account. After going through the unification procedure in §7.3, we consider the boundary value problem

$$\frac{\partial N}{\partial t} = D\Delta N + \lambda\left(1 + \frac{a}{\varepsilon}N_{t-1} - \frac{1}{\varepsilon}N_{t-h}\right)N, \qquad \left|\frac{\partial N}{\partial \nu}\right|_{\partial\Omega} = 0. \tag{9.37}$$

For insects with a pronounced seasonal life cycle the spatially heterogeneous regimes of the boundary value problem (9.37) must be of the following character. The function $N(t, x)$ increases sharply at roughly the same time at all points of the planar region Ω, and then a strong drop occurs in a time close to h. The spatial heterogeneity is expressed mainly in that the density spikes differ markedly in height in different local habitat areas. This implies that the mechanism referred to in Theorem 9.1 for loss of stability of a homogeneous cycle is not at all similar to that described in §3.2. In our opinion, the situation here is closer to that in §6.5, where the subject was stability of a homogeneous cycle of the mathematical model of the Belousov reaction in an environment with diffusion. We thus assume that, for the boundary value problem linearized on a homogeneous cycle, instability appears first at the lower modes, and there is a shift in the direction of higher modes as D decreases, with recovery of stability at the low modes. In the framework of the nonlinear boundary value problem (9.37) this means that the spatial heterogeneity becomes more and more fine-grained as D decreases. A thorough investigation of all these questions is of considerable interest.

Thus, according to our ideas, for small values of ε the mathematical model of the dynamics of insect populations is a representative of complex physical systems, at least in a certain range of variation of the initial conditions. However, for moderate values of ε the diffusion instability of a homogeneous cycle certainly tends to have a Hutchinson character. This is because the boundary value problem (9.37) can be used also for modelling the dynamics of insect populations with life cycle unconnected with the change of seasons. But then the variation in their density in space and in time must bear the character described in §4. This means that in another range of parameters (9.37) is a biological system.

10. The existence of strongly unstable relaxation cycles

10.1. Relaxation cycles with a complex time ordering. We assume now that in equation (7.4) the integer H_0 is at least 2; that is, we consider insects with a life cycle of two or more years. Under this condition the relaxation cycles constructed in Theorem 7.1 mean biologically that finite spikes in the imago density occur across an H_0-year period. We pose the question of relaxation cycles of equation (7.4) whose existence and stability mean in biological language that the finite spikes in the imago density are observed more often, for example, every year. It turns out that the theorem on the existence of this type of relaxation cycle is valid, but the proof is unconnected with the construction of invariant sets, and brings to mind hunting for duck-cycles of relaxation systems of ordinary differential equations [24]. It is important that also in our case these cycles have the property of strong instability; that is, the variational equations corresponding to them have multipliers with moduli that grow unboundedly as $\varepsilon \to 0$, and (which should be underscored) information about the subtle asymptotics of the cycles is not needed in the proof.

The biological meaning of these facts is clear: in an energetically overexcited system the cycles having an extreme time ordering are unstable.

The applied value of the foregoing is that we have an explanation for the loss of stability of the corresponding cycles of equation (7.11) that are numerically observable for intermediate values of the parameter a and for small h.

10.2. The mathematical aspect of the problem of existence of cycles with complex time ordering. We assume that in equation (7.4)

$$h_0 < 1/2. \tag{10.1}$$

Under this restriction the question of the existence of the relaxation cycles of interest to us is connected with the presence of exponentially stable cycles $M_0, \dots, M_k$ for the one-dimensional mapping

$$M \to M \exp\{r_0 - [1 - \exp(-r_0 h_0)]M\}, \tag{10.2}$$

which differs from (7.39) only in the absence of the factor H_0 multiplying r_0. But on the other hand, we now assume that $k \le H_0 - 2$; that is, for $H_0 = 2$ only an exponentially stable equilibrium state is involved, for $H_0 = 3$ a cycle of period two is allowed, and so on. The meaning of these restrictions will be made clear later.

For simplicity of the exposition we confine ourselves to the case $H_0 = 2$. Below it is technically simpler to work with the unified equation (7.11), in which now $h < 1/3$ in view of (10.1). Further, it is convenient to rewrite the mapping (10.2) in the form

$$z \to (1+h)/2 + z - [\exp \lambda(z+h) - \exp \lambda z]/\lambda, \tag{10.3}$$

assuming that it has an exponentially stable equilibrium state

$$z_0 = \frac{1}{\lambda} \ln \frac{\lambda(1+h)}{2(\exp \lambda h - 1)}. \tag{10.4}$$

Denote by S_1 the collection of nonnegative functions $N_1(t, \varepsilon)$ defined for $-1 \le t \le -(1+h)/2 + \varepsilon\nu_0$, where $\nu_0 > 0$ is a constant. We suppose that each of them satisfies the first two conditions in the definition of the set S_0 of functions introduced in the proof of Theorem 7.1. The third condition now takes the following form: for $-1 + \delta(\varepsilon) \le t \le -3/4$

$$N_1(t, \varepsilon) \le \varepsilon \exp[-\nu(1 - \delta(\varepsilon) + t)/\varepsilon],$$

and for $-3/4 \le t \le (1+h)/2 - \varepsilon\nu_0$

$$N_1(t, \varepsilon) \le \varepsilon \exp[-\nu((1+h)/2 + \varepsilon\nu_0 + t)/\varepsilon],$$

where $\nu > 0$ is a universal constant. Instead of the fourth condition we assume that

$$N_1(t, \varepsilon) \le C_0\varepsilon, \qquad |t + (1+h)/2| \le \varepsilon\nu_0,$$

where $C_0 > 1$ is also a universal constant. The fifth condition is preserved, but of course z_0 is now the number (10.4).

Let S_2 be the set of nonnegative functions $N_2(t, \varepsilon)$, $-h \le t \le 0$, with

$$N_2(t, \varepsilon) \le \varepsilon \exp(\nu t/\varepsilon), \qquad N_2(0, \varepsilon) = \varepsilon.$$

Below we fix a positive number T_{in} such that $|T_{\text{in}} - (1+h)/2| \le \varepsilon\nu_0$. The collection of these numbers is denoted by T_0.

We take functions $N_1(t, \varepsilon)$ and $N_2(t, \varepsilon)$ and a number $T_{\rm in}$ in the classes introduced above, and we construct a solution $N(t, \varepsilon)$ of equation (7.11) by steps for $0 \le t \le 1 - T_{\rm in}$. It will be assumed that for $-T_{\rm in} < t < -h$ the initial function is given by

$$N_{\rm init}(t, \varepsilon) = N(T_{\rm in} + t, \varepsilon); \tag{10.5}$$

that is, in the construction of $N(t, \varepsilon)$ we use $N_1(t, \varepsilon)$ only for $-1 \le t \le -T_{\rm in}$. The initial function is now given on the whole interval $[-1, 0]$ with the help of (10.5). This enables us to extend the solution to a moment of time $T_{\rm out}$ that is asymptotically close to $(1+h)/2$ and such that $N(T_{\rm out}, \varepsilon) = \varepsilon$.

Thus, the problem of the existence of a half-period relaxation cycle has been reduced to the solvability of three equations. The first of them is

$$T_{\rm in} = T_{\rm out}(N_1, N_2, T_{\rm in}, \varepsilon). \tag{10.6}$$

The second has the form

$$N_1(t, \varepsilon) = \Pi_1(N_1, N_2, T_{\rm in}, \varepsilon), \tag{10.7}$$

where $\Pi_1 = N(T_{\rm out} + t, \varepsilon)$ and $-1 \le t \le -(1+h)/2 + \varepsilon\nu_0$. Finally, the third is

$$N_2(t, \varepsilon) = \Pi_2(N_1, N_2, T_{\rm in}, \varepsilon), \tag{10.8}$$

where $\Pi_2 = N(T_{\rm out} + t, \varepsilon)$, $-h \le t \le 0$. The right-hand sides of (10.6)–(10.8) give rise to the quasi-first-return operator Π, which will be considered in the space

$$R \times L(-1, -(1+h)/2 + \varepsilon\nu_0) \times C(-h, 0).$$

THEOREM 10.1. *For $h < 1/3$ and $\lambda(1+h) < 4$ there exist constants ν_0, ν, and C_0 and an ε_0 such that for $0 < \varepsilon < \varepsilon_0$ the quasi-first-return operator Π of (7.11) is defined on the closed convex set $T_0 \times S_1 \times S_2$ and maps this set into itself.*

Of course, the operator Π is not compact, but it is condensing [64]. Therefore, under the conditions of Theorem 10.1 equation (7.11) has a half-period relaxation cycle.

We do not dwell on the proof of Theorem 10.1.

THEOREM 10.2. *The variational equation of each half-period relaxation cycle existing under the conditions of Theorem 10.1 has a multiplier with modulus of order $1/\varepsilon$ as $\varepsilon \to 0$.*

The proof of Theorem 10.2 is also omitted. The scheme of proof of this kind of statement will be presented in the next chapter.

We return to the restriction $k \le H_0 - 2$ formulated at the beginning of the subsection. Here our concern is as follows. We shall integrate equation (7.11) by steps with some initial condition supposedly corresponding to, say, a cycle with three spikes on an interval of time of order $1+h$. Dividing equation (7.11) by $N(t, \varepsilon)$ and integrating the result over suitable time intervals, we see that the assumption of the existence of the indicated cycle implies the corresponding time ordering of it. In particular, the fact that it is strongly unstable becomes clear in this way: small differences in the initial condition which violate the necessary time ordering grow sharply for $t > 0$.

CHAPTER III

Relaxation Cycles of Finite Amplitude in a Mathematical Model of the Simplest Controllable Biocenosis

11. Statement of the problem and two existence theorems for relaxation cycles

11.1. Statement of the problem. The goal of this chapter is to analyze questions of existence, asymptotics, and stability of relaxation cycles of finite amplitude for a system with delay

$$\dot{x} = 1 - x\left(1 + \frac{\alpha}{\varepsilon}y\right), \tag{11.1}$$

$$\dot{y} = \frac{y}{\varepsilon}[x - y(t-h)] - \beta y, \tag{11.2}$$

where the positive parameters α, β, and h have order 1, and the positive parameter ε is assumed to be small. If $x \equiv 1$ and $\beta = 0$ in (11.2), then up to a change of notation we obtain the Hutchinson equation (2.1) for the density of a strongly fertile population. If $x \equiv 1$ and $\beta \neq 0$, then it describes a situation (see equation (1.11)) when there is hunting with constant intensity. In considering the system (11.1), (11.2) it is convenient to assume that x is the prey density and y the predator density. It is assumed that the prey is not capable of maintaining itself (the term $-x$ on the right-hand side of (11.1)), and hence its population must be replenished with constant intensity (the first term on the right-hand side of (11.1)). Finally, the term $-\alpha\varepsilon^{-1}xy$ in (11.1) means that the predator strongly oppresses the prey.

Thus, the system (11.1), (11.2) is the simplest model for the functioning of an artificial biocenosis.

On the other hand, the system (11.1), (11.2) has already been used long enough (roughly since the mid-1960's) to model the dynamics of quantum generators (in the physical interpretation of it the variables x and y must be divided by ε).

As we shall see in what follows, the presence or absence of relaxation cycles for the system (11.1), (11.2) depends in a very essential way on the behavior of the function

$$f(\sigma) = F(\sigma, 1 - \exp(-\sigma)) \tag{11.3}$$

on the positive semiaxis, where

$$F(\sigma, x) = \left(1 - \frac{1}{\alpha\beta}\right)\sigma - 1 + \exp(-\sigma) + \frac{1}{\alpha\beta}\left(\frac{1}{\beta} - x\right)[1 - \exp(-\beta\sigma)]. \tag{11.4}$$

It follows from (11.4) that $\sigma = 0$ is a double zero of it. But since the function (11.3) is a solution of some fifth-order linear differential equation with real roots of the characteristic polynomial, there are at most four of them. Therefore, it is convenient to distinguish the following four cases (we do not dwell on the elementary proofs).

FIRST CASE. The function (11.3) is negative for $\sigma > 0$ if

$$\alpha\beta < 1, \qquad \alpha < 3. \tag{11.5}$$

SECOND CASE. The function (11.3) has a unique positive zero σ_0 with $f'(\sigma_0) > 0$ if

$$\alpha\beta > 1, \qquad \alpha < 3. \tag{11.6}$$

THIRD CASE. The same as in the second case except with $f'(\sigma_0) < 0$ if

$$\alpha\beta < 1, \qquad \alpha > 3. \tag{11.7}$$

FOURTH CASE. The function (11.3) is positive for $\sigma > 0$ if

$$\alpha\beta > 1, \qquad \alpha > 3. \tag{11.8}$$

When the inequalities (11.5) hold, the predator is not very voracious, and it is not hunted with extreme intensity, that is, the first case is the simplest from a biological point of view. The second case has the most applied significance, since as before the predator is not too voracious, but the hunting can be of any intensity. The third and fourth cases relate to the situation when the predator is fairly voracious.

The second case is also of greatest interest for quantum physics, as follows from the typical parameter values given in [15], which become the following in our notation:

$$\alpha = 1\,,\, 2.5\,,\, 0.611\,; \qquad \beta = 2\,,\, 0.952\,,\, 36\,,\, 364.$$

Below we give our main attention to the second case. We discuss separately the possibility of extending some of the results to the first and third cases. The fourth case will not be touched upon, since in general the inequalities (11.8) prevent the existence of a relaxation cycle of finite amplitude.

11.2. Two lemmas relating to the second case with $\sigma_0 > h$. Let

$$\varphi(\sigma) = \Phi[\sigma\,,\, 1 - \exp(-\sigma)], \tag{11.9}$$

where

$$\Phi(\sigma\,,\, x) = F(h\,,\, x) + \sigma - h - \exp(-h) + \exp(-\sigma). \tag{11.10}$$

Under the conditions in the subsection title,

$$\varphi(h) = f(h) < 0. \tag{11.11}$$

Therefore, as σ increases, the function (11.9) has a zero $\sigma_1 > h$ that is obviously simple.

LEMMA 11.1. $\sigma_1 < \sigma_0$.

If the opposite holds, that is, if $\sigma_1 \geq \sigma_0$, then we introduce the function $\psi(\sigma) = \varphi(\sigma) - f(\sigma)$. According to (11.11), $\psi(h) = 0$, and

$$\psi'(h) = \frac{1}{\alpha\beta}[1 - \exp(-\beta h)] + \frac{1}{\alpha}[1 - \exp(-h)]\exp(-\beta h) > 0.$$

Further, $\psi(\sigma_1) \leq 0$ by the foregoing, and hence there exists an $h < \sigma_* \leq \sigma_1$ such that $\psi(\sigma_*) = 0$ and $\psi'(\sigma_*) \leq 0$. The latter is impossible. Indeed,

$$\psi'(\sigma_*) = \frac{1}{\alpha\beta}[1 - \exp(-\beta\sigma_*)] + \frac{1}{\alpha}[1 - \exp(-\sigma_*)]\exp(-\beta\sigma_*)$$
$$+ \frac{1}{\alpha\beta}[\exp(-\beta h) - \exp(-\beta\sigma_*)]\exp(-\sigma_*) > 0,$$

since $\sigma_* > h$.

We touch on the question of how σ_1 depends on h.

LEMMA 11.2. $\sigma_1'(h) > 0$.

For a proof of this it is convenient to denote the right-hand side of (11.10) by $\Phi(h, \sigma, x)$, explicitly underscoring the dependence on h. We have

$$\sigma_1'(h) = -\Phi_h/[\Phi_\sigma + \Phi_x \exp(-\sigma_1)]. \tag{11.12}$$

But

$$\varphi'(\sigma_1) = \Phi_\sigma + \Phi_x \exp(-\sigma_1) > 0. \tag{11.13}$$

Therefore, we must show that

$$\Phi_h < 0. \tag{11.14}$$

We note that in (11.12)–(11.14) the derivatives are taken for $\sigma = \sigma_1$ and $x = 1 - \exp(-\sigma_1)$. It follows from (11.4) and (11.10) that

$$\Phi_h = -\frac{1}{\alpha\beta}[1 - \exp(-\beta h)] - \frac{1}{\alpha}[1 - \exp(-\sigma_1)]\exp(-\beta h);$$

that is, we get the required inequality (11.14).

11.3. Main theorem on the existence of relaxation cycles. In what follows, $R \times C(-h, 0)$ is the phase space of the system (11.1), (11.2). Denote by $x(t, \varepsilon)$, $y(t, \varepsilon)$ its solution with the initial conditions

$$x(0, \varepsilon) = x > 0, \tag{11.15}$$

$$0 \leq y(t, \varepsilon) \leq \varepsilon \exp\frac{m}{\varepsilon}t, \qquad y(0, \varepsilon) = \varepsilon, \tag{11.16}$$

where $m > 0$ is a constant. It follows from (11.16) that with exponential accuracy equation (11.2) can be approximated by the usual equation

$$\dot{y} = \frac{1}{\varepsilon}xy - \beta y \tag{11.17}$$

for $0 \leq t \leq h - \delta$, where along with ε the constant $\delta > 0$ can be arbitrarily decreased. We multiply (11.17) by α and add the result to (11.1). As a result, instead of (11.1), (11.17) we obtain the system

$$\dot{z} = 1 - x - \beta(z - x), \tag{11.18}$$

$$\varepsilon\dot{x} = \varepsilon(1 - x) - x(z - x), \tag{11.19}$$

where

$$z = x + \alpha y. \tag{11.20}$$

In the language of biology the choice of the initial conditions (11.15), (11.16) means that for $0 \le t \le \delta$ there is a sharp spike in the predator density, while the prey density drops sharply, that is,

$$\lim_{\delta\to 0}\lim_{\varepsilon\to 0} x(\delta, \varepsilon) = 0. \tag{11.21}$$

However, in view of (11.18) the function (11.20) is almost without change on this time interval. Therefore,

$$\lim_{\delta\to 0}\lim_{\varepsilon\to 0} y(\delta, \varepsilon) = x/\alpha. \tag{11.22}$$

We deduce from (11.18)–(11.22) that the zero asymptotics for $\delta \le t \le h-\delta$ has the form

$$x_0(t) \equiv 0, \quad y_0(t) = \frac{1}{\alpha\beta} - \frac{1}{\alpha}\left(\frac{1}{\beta} - x\right)\exp(-\beta t). \tag{11.23}$$

For $h+\delta \le t \le 2h-\delta$ the delayed feedback operates, and this leads to a drop in the density of the predator and an increase in the density of the prey. It follows from (11.23) and (11.1), (11.2) that on the given time interval the following formulas hold for the zero asymptotics of the solutions $x(t, \varepsilon)$, $y(t, \varepsilon)$:

$$x_0(t) = 1 - \exp[-(t-h)], \quad y_0(t) \equiv 0. \tag{11.24}$$

Next, in view of (11.23) and (11.2) the solution component $y(t, \varepsilon)$ is exponentially small for the indicated t, with the exponent of order $1/\varepsilon$ and the coefficient equal to $F(\sigma, x)$, $\sigma = t - h$, up to small quantities, where F is the function (11.4). Of course, it is assumed that for this solution

$$F(\sigma, x) < 0, \qquad 0 < \sigma \le h. \tag{11.25}$$

The function $y(t, \varepsilon)$ thus remains exponentially small also for certain $t \ge 2h + \delta$, and the corresponding coefficient is equal to $\Phi(\sigma, x)$ up to small quantities, where Φ is the function (11.10). Therefore, there is a first $\sigma(x)$ such that $\Phi(\sigma(x), x) = 0$. Hence, by (11.24), the condition

$$x = 1 - \exp[-\sigma(x)] \tag{11.26}$$

is necessary for the existence of a relaxation cycle of finite amplitude for the system (11.1), (11.2).

THEOREM 11.1. *Under the conditions of the second case suppose that*

$$\sigma_0 > h. \tag{11.27}$$

Then there is an $\varepsilon_0 > 0$ *such that for* $0 < \varepsilon < \varepsilon_0$ *the system* (1.1), (1.2) *has a relaxation cycle* $x(t, \varepsilon)$, $y(t, \varepsilon)$, *and*

$$\lim_{\varepsilon\to 0} x(0, \varepsilon) = x_1, \qquad \lim_{\varepsilon\to 0} T(\varepsilon) = h + \sigma_1,$$

where $x_1 = 1 - \exp(-\sigma_1)$, *and* $T(\varepsilon)$ *is its period.*

PROOF. We begin with the question of the implicit function $\sigma = \sigma(x)$, $\sigma(x_1) = \sigma_1$, determined from the equality $\Phi(\sigma, x) = 0$. In (11.13) we have $\Phi_x < 0$; therefore, $\Phi_\sigma > 0$. Consequently, the smooth function $\sigma(x)$ is uniquely determined, and

$$\sigma'(x_1) = -\left.\frac{\Phi_x}{\Phi_\sigma}\right|_{x=x_1,\, \sigma=\sigma_1} > 0. \tag{11.28}$$

We now introduce the one-dimensional mapping

(11.29) $$x \to 1 - \exp[-\sigma(x)].$$

By construction, it is defined in a neighborhood of its unique fixed point x_1. We show that

(11.30) $$0 < \sigma'(x_1)\exp[-\sigma(x_1)] < 1.$$

Indeed, dividing (11.13) by $-\Phi_x$ and using (11.28), we arrive at (11.30).

To conclude the proof we consider the closed convex set of initial conditions (11.15), (11.16), assuming that x is in a sufficiently small neighborhood of x_1, and the constant m is suitably decreased. The first-return operator, which is defined in the obvious way on this set, transforms it into itself: just observe that the first component of the operator depends weakly on the second, and invoke (11.30).

It follows from the proof of the theorem that the second component of the first-return operator tends to zero as $\varepsilon \to 0$ uniformly with respect to the initial conditions under consideration, while the first component converges to the mapping (11.29). This means that a version of the C-convergence theorem holds. The question of the validity of the C^1-convergence theorem will be treated in §13, where its treatment coincides completely with the case of ordinary differential equations.

The statement of Theorem 11.1 is preserved without additional restrictions under the conditions of the first case. For it to be valid in the third case we must replace the inequality (11.27) by the strict reverse inequality and require in addition that

$$F[\sigma, 1 - \exp(-\sigma_1)] < 0, \qquad 0 < \sigma \le h.$$

We remark that there is an analogue of our Theorem 11.1 in [15], where an implicitly given one-dimensional mapping whose properties are not studied appears in the formulation. The question of stability of the relaxation cycle, which is a part of the C^1-convergence theorem, is also said in [15], but nothing at all is said about how to investigate it.

11.4. A theorem on the existence of relaxation cycles with short periods. The cycles of the system (11.1), (11.2) with period less than h are such cycles. Before formulating the basic assertion relating to the question of their existence, we introduce some quantities. In (11.10) we replace h by h_k, and h will be taken to be a background parameter. Then we consider the equations

(11.31) $$k\sigma_1(h_k) = h - (k+1)h_k, \qquad k = 1, 2, \dots,$$

with respect to h_k. By Lemma 11.2, each of them is uniquely solvable, and $h_k < h/(2k+1)$.

THEOREM 11.2. *Under the conditions of the second case suppose that that for an arbitrarily fixed* k

(11.32) $$\sigma_0 > h_k.$$

Then there is an $\varepsilon_k > 0$ *such that for* $0 < \varepsilon < \varepsilon_k$ *the system* (11.1), (11.2) *has a relaxation cycle such that*

(11.33) $$\lim_{\varepsilon \to 0} T_k(\varepsilon) = (h - h_k)/k,$$

where $T_k(\varepsilon)$ *is its period.*

PROOF. In equation (11.2) we replace the delay h by $h = h - kT$. By (11.32), the parameter T can be taken close enough to the right-hand side of (11.33) that $h < \sigma_0$. But then by Theorem 11.1 the system (11.1), (11.2), modified as indicated above, has a relaxation cycle with period

$$\tilde{T} = \tilde{h} + \sigma_1(\tilde{h}) + o(1). \tag{11.34}$$

We require that $\tilde{T} = T$. It follows from (11.34) and Lemma 11.2 that this equation is solvable, and in view of (11.31) the limit equality (11.33) holds for the solution $T_k(\varepsilon)$ of it. The proof is finished, since a T-periodic solution of the modified system is a T-periodic solution of the original system (11.2), (11.2).

Theorem 11.2 is outwardly close to Theorem 10.1. But in fact there is an essential difference, connected with the circumstance that Theorem 10.1 is concerned with cycles with period given in a fairly inflexible way for biological reasons. It is in this connection that the device used in the proof of Theorem 11.2 does not work under the conditions of Theorem 10.1.

We now describe a method to be used in §13 to solve the problem of stability of relaxation cycles with short periods.

We begin by making the normalization substitution $v_1 = v$, $w_1 = w/y$ in the variational system

$$\dot{v} = -\left(1 + \frac{\alpha}{\varepsilon}y\right)v - \frac{\alpha}{\varepsilon}xw, \tag{11.35}$$

$$\dot{w} = \frac{1}{\varepsilon}yv + \frac{\dot{y}}{y}w - \frac{1}{\varepsilon}yw(t-h). \tag{11.36}$$

We get

$$\dot{v} = -\left(1 + \frac{\alpha}{\varepsilon}y\right)v - \frac{\alpha}{\varepsilon}xyw, \tag{11.37}$$

$$\dot{w} = \frac{1}{\varepsilon}v - \frac{1}{\varepsilon}y(t-h, \varepsilon)w(t-h), \tag{11.38}$$

where the indices are omitted. Denote by

$$v(t, \varepsilon)\exp\mu t, \quad w(t, \varepsilon)\exp\mu t \tag{11.39}$$

an arbitrary Lyapunov-Floquet solution of it, that is, the functions v, w are periodic in t with period T. We substitute the functions (11.39) in (11.37), (11.38) and write the result in the form

$$\dot{v} + \mu v = -\left(1 + \frac{\alpha}{\varepsilon}y\right)v - \frac{\alpha}{\varepsilon}xyw,$$

$$\dot{w} + \mu w = \frac{1}{\varepsilon}v - \frac{1}{\varepsilon}y(t-\tilde{h}, \varepsilon)w(t-\tilde{h})\exp(-\mu h),$$

where $\tilde{h} = h - kT$. We point out that $w(t-h)$ is written instead of $w(t-h, \varepsilon)$; that is, we regard it as an independent system. Making the substitutions

$$v = v_1\exp(-\mu t), \qquad w = w_1\exp(-\mu t)$$

and again omitting the indices, we arrive at the system

$$\dot{v} = -\left(1 + \frac{\alpha}{\varepsilon}y\right)v - \frac{\alpha}{\varepsilon}xyw, \tag{11.40}$$

$$\dot{w} = \frac{1}{\varepsilon}v - \frac{A}{\varepsilon}y(t-\tilde{h}, \varepsilon)w(t-\tilde{h}), \tag{11.41}$$

where $A = \exp(-\mu kT)$.

The system (11.40), (11.41) is nice in that, in particular, to investigate it we need know only the asymptotics of the relaxation cycle in Theorem 11.1. For an analysis of this system we assume that A is an independent complex parameter. Under this condition the limits of all the multipliers of (11.40), (11.41) as $\varepsilon \to 0$ will be found. Let $\nu(A)$ be one of them. We consider the following equation in A:

$$A^{1/k}\nu(A) = 1. \tag{11.42}$$

We derive the following criterion from the connection described above between the system (11.37), (11.38) and the system (11.40), (11.41). If each solution $A \neq 1$ of each of the equations (11.42) lies outside the unit disk and if 1 is a simple solution, then the corresponding relaxation cycle with short period is orbitally exponentially stable. But if one of the equations (11.42) has a solution A with $|A| < 1$, then the cycle is unstable.

12. Asymptotics of relaxation oscillations

12.1. A description of the scheme of investigation. It is natural that, as in the case of ordinary equations [9], the structure of the whole asymptotic expression of the relaxation cycle referred to in Theorem 11.1 for the system (11.1), (11.2) is determined by the asymptotic expression on the rapid motion part, that is, by the asymptotic expression for $x(t, \varepsilon)$, $y(t, \varepsilon)$ in a suitable asymptotically small neighborhood of $t = 0$. However, a new point also arises: for the asymptotic expression to have the simplest form we must choose a suitable phase of the cycle. As will be clear from what follows, in our case it is more convenient to construct the asymptotics of the periodic solution of (11.1), (11.2) with initial conditions

$$x(0, \varepsilon) = \frac{q}{2}, \quad 0 \leq y(t, \varepsilon) \leq \frac{q}{2\alpha} \exp\frac{mt}{\varepsilon}, \quad y(0, \varepsilon) = \frac{q}{2\alpha}, \tag{12.1}$$

with $m > 0$ and q a constant in a sufficiently small neighborhood of x_1, whose components are also denoted by $x(t, \varepsilon)$, $y(t, \varepsilon)$. It will follow from what is presented below that this periodic solution differs by an asymptotically small phase (of order $\varepsilon \ln(1/\varepsilon)$) from the cycle supplied in Theorem 11.1.

Thus, by virtue of (12.1) we begin with asymptotic integration of the system (11.18), (11.19), supplemented by the initial conditions

$$x(0, \varepsilon) = q/2, \quad z(0, \varepsilon) = q. \tag{12.2}$$

After we are done with an asymptotically small neighborhood of $t = 0$, we construct the asymptotics of the problem (11.18), (11.19), (12.2) up to an asymptotically small neighborhood of $t = h$. Then we invoke the system (11.1), (11.2) and construct the asymptotics of solutions for large values of t. Finally, from the equalities $x(T, \varepsilon) = q/2$ and $y(T, \varepsilon) = q/2\alpha$ we find the period $T > 2h$ of the relaxation cycle and the constant $q(\varepsilon)$ in (12.1), (12.2).

In implementing this program we do not dwell on the standard proof of the asymptotic formulas, referring, for example, to [31] and [62] (see also §8, where a way of proving them was described in a much more complex case).

12.2. The rapid motion part. We begin by constructing the asymptotics of the Cauchy problem (11.18), (11.19), (12.2) for $-\varepsilon^{\lambda} \leq t \leq \varepsilon^{\lambda}$, $0 < \lambda < 1$.

We remark that the system (11.18), (11.19) has an exponentially stable (with exponent of order ε^{-1}) invariant integral manifold $x = \varphi(z, \varepsilon)$, where φ is a jointly smooth function of its variables, and

$$\varphi(z, \varepsilon) = \varepsilon/z + O(\varepsilon^2). \tag{12.3}$$

Therefore, from the point with coordinates (12.2) there is a "drop" onto the integral manifold (12.3) for $t > 0$ which can be described with the help of the technique in [62] (for $t < 0$ there is a drop onto the unstable manifold $x = z$).

It follows from the foregoing [62] that in (11.18), (11.19) it is necessary to make the substitution $t = \varepsilon\tau$ and then, in the resulting system

$$\dot{z} = \varepsilon[1 - x - \beta(z - x)], \qquad \dot{x} = \varepsilon(1 - x) - x(z - x), \tag{12.4}$$

to set

$$z = q + \sum_{k=1}^{\infty} \varepsilon^k z_k(\tau), \qquad x = x_0(\tau) + \sum_{k=1}^{\infty} \varepsilon^k x_k(\tau), \tag{12.5}$$

where $x_0(\tau) = q/(1 + \exp q\tau)$. Equating coefficients of like powers of ε, we arrive at the recursion sequence of differential equations

$$\dot{z}_k = F_k(\tau), \qquad \dot{x}_k = (2x_0(\tau) - q)x_k + G_k(\tau), \tag{12.6}$$

where F_k depends on the x_n and z_m with indices $n, m \leq k - 1$, and G_k on the x_n and z_m with indices $n \leq k - 1$ and $m \leq k$. Supplementing (12.6) by the initial conditions $x_k(0) = z_k(0) = 0$ (see (12.2)), we determine the coefficients in the expansions (12.5) successively. For example,

$$z_1(\tau) = (1 - q)\tau + (\beta - 1)\ln 2 - (\beta - 1)\ln(1 + \exp q\tau), \tag{12.7}$$

$$x_1(\tau) = \int_0^\tau [1 - x_0(s) - z_1(s)x_0(s)] \exp\left\{\int_s^\tau (2x_0(\sigma) - q)\, d\sigma\right\} ds. \tag{12.8}$$

It turns out that the following representations hold (as is easily proved by induction):

$$x_k(\tau) = P_{k,+}^{k-1}(\tau) + \omega_{k,+}(\tau), \qquad z_k(\tau) = Q_{k,+}^{k}(\tau) + \delta_{k,+}(\tau) \tag{12.9}$$

as $\tau \to \infty$;

$$x_k(\tau) = P_{k,-}^{k}(\tau) + \omega_{k,-}(\tau), \qquad z_k(\tau) = Q_{k,-}^{k}(\tau) + \delta_{k,-}(\tau) \tag{12.10}$$

as $\tau \to -\infty$. Here $P_{k,+}^m$, $Q_{k,+}^m$, $P_{k,-}^m$, and $Q_{k,-}^m$ are mth-degree polynomials, and the remainders $\omega_{k,+}$, $\delta_{k,+}$ and $\omega_{k,-}$, $\delta_{k,-}$ tend to zero at an exponential rate as $\tau \to \infty$ and $-\infty$, respectively.

We remark that as $\tau \to -\infty$ the solution of the Cauchy problem (12.4), (12.2) approaches the integral manifold $x = z$ at an exponential rate. Therefore, $P_{k,-}^k(\tau) = Q_{k,-}^k(\tau)$ for all $k \geq 1$.

Using the representations (12.9), (12.10), we can construct the asymptotics of the solution of the Cauchy problem (12.4), (12.2) with any degree of accuracy. However, for our purposes it is enough to have the equalities

$$x(\tau, \varepsilon) = x_0(\tau) + \varepsilon x_1(\tau) + O(\varepsilon^{2\lambda}), \tag{12.11}$$

$$z(\tau, \varepsilon) = q + \varepsilon z_1(\tau) + O(\varepsilon^{2\lambda}), \tag{12.12}$$

where $-\varepsilon^{\lambda-1} \leq \tau \leq \varepsilon^{\lambda-1}$, $1/2 < \lambda < 1$, $x_1(\tau)$ and $z_1(\tau)$ are the functions (12.7) and (12.8), and the remainders are uniform with respect to the indicated values of τ.

12.3. The slow motion part. We recall that for $\varepsilon^\lambda \le t \le h - \varepsilon^\lambda$, $0 < \lambda < 1$, the solution of the Cauchy problem (11.18), (11.19), (12.2) lies on the integral manifold (12.3). Therefore, the problem reduces to computing the asymptotics of the initial condition $z(0, \varepsilon)$ for the equation

$$\dot{z} = 1 - \varphi(z, \varepsilon) - \beta(z - \varphi(z, \varepsilon)) \tag{12.13}$$

on the integral manifold (12.3).

We substitute $\tau = t/\varepsilon$ in (12.5), use the representations (12.9), discard the exponentially small terms, and then set $t = 0$. This gives us the asymptotic representation

$$z(0, \varepsilon) = q + \sum_{k=1}^{\infty} \varepsilon^k Q^k_{k,+}(0), \tag{12.14}$$

where, by (12.7),

$$Q^1_{1,+}(0) = (\beta - 1)\ln 2. \tag{12.15}$$

From (12.3) and (12.13)–(12.15) we have the asymptotic formulas

$$z(t, \varepsilon) = z_0(t) + \varepsilon z_1(t) + O(\varepsilon^2), \tag{12.16}$$

$$x(t, \varepsilon) = \varepsilon / z_0(t) + O(\varepsilon^2), \tag{12.17}$$

where

$$z_0(t) = \frac{1}{\beta}(1 - \exp(-\beta t)) + q\exp(-\beta t), \tag{12.18}$$

(12.19)

$$\begin{aligned} z_1(t) &= (1 - \beta)(1 + \ln 2\beta q)\exp(-\beta t) \\ &\quad + (1 - \beta)\{\beta(\beta q - 1)t - \exp \beta t - \ln(1 + (\beta q - 1)\exp(-\beta t))\}\exp(-\beta t), \end{aligned}$$

and the remainders have the indicated orders uniformly with respect to $\varepsilon^\lambda \le t \le h - \varepsilon^\lambda$, $0 < \lambda < 1$.

12.4. Construction of the asymptotics for $t \ge h - \varepsilon^\lambda$. For these t we are dealing with the system

$$\dot{z} = 1 - x - \beta(z - x) - \frac{1}{\alpha\varepsilon}(z - x)(z(t - h) - x(t - h)), \tag{12.20}$$

$$\varepsilon\dot{x} = \varepsilon(1 - x) - x(z - x), \tag{12.21}$$

and we keep the notation $x(t, \varepsilon)$, $z(t, \varepsilon)$ for the solution of it with initial conditions $x(h - \varepsilon^\lambda, \varepsilon)$, $z(t, \varepsilon)$, $-\varepsilon^\lambda \le t \le h - \varepsilon^\lambda$. It follows from (12.20) that the asymptotic behavior of $x(t, \varepsilon)$, $z(t, \varepsilon)$ is determined by the asymptotics of $z(t - h, \varepsilon) - x(t - h, \varepsilon)$, which has already been constructed. Therefore, we present the final formulas at once.

For $h - \varepsilon^\lambda \le t \le h + \varepsilon^\lambda$ we make the substitution $t = \varepsilon\tau + h$ in (12.20), (12.21) and let $x(\tau, \varepsilon) = x(t, \varepsilon)|_{t=\varepsilon\tau+h}$, $z(\tau, \varepsilon) = z(t, \varepsilon)|_{t=\varepsilon\tau+h}$. Then, uniformly with respect to $-\varepsilon^{\lambda-1} \le \tau \le \varepsilon^{\lambda-1}$, we have the asymptotic equalities

$$z(\tau, \varepsilon) = a_0(\tau) + \varepsilon a_1(\tau) + O(\varepsilon^{2\lambda}), \tag{12.22}$$

$$x(\tau, \varepsilon) = \varepsilon b_1(\tau) + O(\varepsilon^{2\lambda}), \tag{12.23}$$

where

$$a_0(\tau) = \varkappa_1 \exp\left\{-\int_{-\infty}^{r} \frac{1}{\alpha}(q - x_0(s))\, ds\right\},$$

$$b_1(\tau) = \int_{-\infty}^{\tau} \exp\left\{\int_{\tau}^{s} a_0(\sigma)\, d\sigma\right\} ds,$$

$$a_1(\tau) = \varkappa_2 \exp\left\{-\frac{1}{\alpha}\int_{-\infty}^{\tau} (q - x_0(s))\, ds\right\}$$
$$+ (\mathrm{H})\int_{-\infty}^{\tau} \left[1 - \beta a_0(s) + \frac{1}{\alpha} b_1(s)(q - x_0(s)) - \frac{1}{\alpha} a_0(s)(z_1(s) - x_1(s))\right]$$
$$\times \exp\left\{-\int_{s}^{\tau} \frac{1}{\alpha}(q - x_0(\sigma))\, d\sigma\right\} ds,$$

$\varkappa_1$ and $\varkappa_2$ are the values of the respective functions (12.18) and (12.19) at $t = h$, the functions $x_0(s)$, $x_1(s)$, and $z_1(s)$ are from (12.5), and $(\mathrm{H})\int$ is the Hadamard regularization [62].

For $h + \varepsilon^\lambda \le t \le h + \sigma(q) - \varepsilon^\lambda$, $0 < \lambda < 1$, where $\sigma = \sigma(x)$ is the function introduced in the proof of Theorem 11.1, we have

$$x(t, \varepsilon) = z(t, \varepsilon) = 1 - (1 - \mu\varepsilon)\exp[-(t - h)] + O(\varepsilon^2), \tag{12.24}$$

where

$$\mu = \lim_{\tau\to\infty} (b_1(\tau) - \tau). \tag{12.25}$$

It remains to describe a method for computing the constant $q(\varepsilon)$ in (12.1) and (12.2), and the period $T(\varepsilon)$ of the relaxation cycle. Let $t_0 = t_0(q, \varepsilon)$ be the first zero of the function $x(t, \varepsilon) - q/2$, where $x(t, \varepsilon)$ is the component of the solution of the system (12.20), (12.21) for $t \ge h + \varepsilon^\lambda$. We precede the asymptotic computation of it by the following remarks. First, the complete asymptotics (cf. (12.24)) of $x(t, \varepsilon)$ for $h + \varepsilon^\lambda \le t \le h + \sigma(q) - \varepsilon^\lambda$ has the form

$$x(t, \varepsilon) = 1 - \left(1 + \sum_{k=1}^{\infty} r_k \varepsilon^k\right)\exp[-(t - h)], \tag{12.26}$$

where $r_1 = -\mu$ (see (12.25)). Indeed, for these t the motion takes place along the integral manifold $x = z$ of the system (11.18), (11.19), on which the initial condition $x(h, \varepsilon)$ for the equation $\dot{x} = 1 - x$ is determined according to the rule described in §12.3 from the representation (12.23). Second, we have with exponential accuracy from (11.2) for $2h + \varepsilon^\lambda \le t \le h + \sigma(q) - \varepsilon^\lambda$, $0 < \lambda < 1$, that

$$y(t, \varepsilon) = \frac{q}{2\alpha}\exp\left\{\frac{1}{\varepsilon}\int_0^t x(s, \varepsilon)\, ds - \frac{1}{\varepsilon}\int_{-\varepsilon^\lambda}^{h+\varepsilon^\lambda} y(s, \varepsilon)\, ds - \beta t\right\}. \tag{12.27}$$

Using here the formula (12.26) and the asymptotics of $x(t, \varepsilon)$, $y(t, \varepsilon)$ constructed earlier, we see that

$$y(t, \varepsilon) = \frac{q}{2\alpha}\exp\left\{\frac{1}{\varepsilon}\Phi(t - h, q) - \beta t + \sum_{k=0}^{\infty} \gamma_k \varepsilon^k\right\}, \tag{12.28}$$

where $\Phi = \Phi(\sigma, x)$ is the function (11.10). Finally, for $h + \sigma(q) - \varepsilon^\lambda \le t \le h + \sigma(q) + \varepsilon^\lambda$, $0 < \lambda < 1$, we make the substitution $\tau = (1 - h - \sigma(q))/\varepsilon$ in the system (11.18), (11.19) and arrive at the system (12.4). We find the solutions of the latter in the form of series

$$x = x_0(\tau) + \sum_{k=1}^{\infty} \varepsilon^k x_k(\tau), \qquad z = g(q) + \sum_{k=1}^{\infty} \varepsilon^k z_k(\tau), \tag{12.29}$$

where $g(q) = 1 - \exp[-\sigma(q)]$, and the coefficients x_k and z_k are solutions of equations of the form (12.6) which are chosen from "matching" conditions as $\tau \to -\infty$ with the functions (12.26), (12.28) for $t = h + \sigma(q) - \varepsilon\tau$. For example, $x_0(\tau) = g(q)/(1 + c_0 \exp(g(q)\tau)$, where

$$c_0 = \frac{q}{2g(q)} \exp[\gamma_0 - \beta(h + \sigma(q))],$$

and the constant γ_0 is from (12.28).

Thus, all the necessary facts for computing $t_0(q, \varepsilon)$ and $q(\varepsilon)$ have been given. If we now set the right-hand side of (12.29) equal to $q/2$, we get the asymptotic representation

$$t_0(q, \varepsilon) = h + \sigma(q) + \sum_{k=1}^{\infty} \varepsilon^k t_k(q). \tag{12.30}$$

Therefore, the analogous asymptotic formula is valid for the parameter q determined from the equation $z(\tau, \varepsilon)|_{\tau=(t_0-h-\sigma(q))/\varepsilon} = q$, where $z(\tau, \varepsilon)$ is the series (12.29). Using this formula in (12.30), we deduce that

$$T(\varepsilon) = h + \sigma_1 + \sum_{k=1}^{\infty} \varepsilon^k T_k,$$

and this concludes the construction of the asymptotics of a relaxation cycle of the system (11.1), (11.2).

13. Stability properties of relaxation cycles

13.1. The basic lemma. We recall that the question of stability of the relaxation cycles supplied by Theorems 11.1 and 11.2 for the system (11.1), (11.2) reduces to an analysis of the distribution of the multipliers of the system

$$\dot{v} = -\left(1 + \frac{\alpha}{\varepsilon} y\right) v - \frac{\alpha}{\varepsilon} xyw, \tag{13.1}$$

$$\dot{w} = \frac{1}{\varepsilon} v - \frac{A}{\varepsilon} y(t-h, \varepsilon) w(t-h), \tag{13.2}$$

where A is an arbitrary complex parameter, and $x(t, \varepsilon)$, $y(t, \varepsilon)$ is the trajectory of the relaxation cycle with initial conditions (12.1).

Let $v(t, \varepsilon)$, $w(t, \varepsilon)$ be a solution of the system (13.1), (13.2) with arbitrary initial condition $(v_0, w_0(t))$ in the unit ball about zero in the phase space $R \times C(-h-\delta_0, -\delta_0)$, where $\delta_0 > 0$ is sufficiently small and fixed. We define the functions $\tilde{v}(t, \varepsilon)$, $\tilde{w}(t, \varepsilon)$ by the following rule: for $-\delta_0 \le t \le h - \delta_0$ the functions $\tilde{v}(t, \varepsilon)$ and $\tilde{w}(t, \varepsilon)$ form the solution of the Cauchy problem for the system

$$\dot{v} = -\left(1 + \frac{\alpha}{\varepsilon} y\right) v - \frac{\alpha}{\varepsilon} xyw, \tag{13.3}$$

$$\dot{w} = \frac{1}{\varepsilon} v \tag{13.4}$$

with the initial conditions $\tilde{v}(-\delta_0, \varepsilon) = v_0$ and $\tilde{w}(-\delta, \varepsilon) = w_0$, where $w_0 = w_0(-\delta_0)$; for $t \geq h - \delta_0$ we take $\tilde{v}(t, \varepsilon)$, $\tilde{w}(t, \varepsilon)$ to be the solution of (13.1), (13.2) with initial conditions $\tilde{v}(h - \delta_0, \varepsilon)$ and $\tilde{w}(t, \varepsilon)$, $-\delta_0 \leq t \leq h - \delta_0$.

LEMMA 13.1. *The following asymptotic equalities hold*:

$$v(t, \varepsilon) = \tilde{v}(t, \varepsilon) + O(\exp(-a/\varepsilon)),$$
$$w(t, \varepsilon) = \tilde{w}(t, \varepsilon) + O(\exp(-a/\varepsilon)), \tag{13.5}$$

$a > 0$, *uniformly with respect to* $-\delta_0 \leq t \leq T(\varepsilon) - \delta_0$ *and with respect to the initial conditions* $(v_0, w_0(t))$.

For the proof we remark that the formulas (13.5) follow from general asymptotic properties of $y(t, \varepsilon)$ when $-\delta_0 \leq t \leq h - \delta_0$, and from the method of steps and very coarse estimates of solutions of the system (13.1), (13.2) when $t \geq h - \delta_0$.

13.2. An investigation of stability of the relaxation cycles. Lemma 13.1 reduces the problem of the distribution of the multipliers of (13.1), (13.2) to an analysis of the two-dimensional operator

$$W(v_0, w_0(t), \varepsilon) = (\tilde{v}(T(\varepsilon) - \delta_0, \varepsilon), \ \tilde{w}(t + T(\varepsilon) - \delta_0, \varepsilon)), \tag{13.6}$$

which it is convenient to identify with the two-dimensional matrix

$$\begin{pmatrix} a_{11}(\varepsilon) & a_{12}(\varepsilon) \\ a_{21}(\varepsilon) & a_{22}(\varepsilon) \end{pmatrix}, \tag{13.7}$$

where $a_{11} = \tilde{v}_1(T(\varepsilon) - \delta_0, \varepsilon)$, $a_{21} = \tilde{w}_1(T(\varepsilon) - \delta_0, \varepsilon)$, $a_{12} = \tilde{v}_2(T(\varepsilon) - \delta_0, \varepsilon)$, $a_{22} = \tilde{w}_2(T(\varepsilon) - \delta_0, \varepsilon)$, and $(\tilde{v}_1, \tilde{w}_1)$ and $(\tilde{v}_2, \tilde{w}_2)$ are the functions $\tilde{v}$ and $\tilde{w}$ for $v_0 = 1$, $w_0 = 0$ and $v_0 = 0$, $w_0 = 1$, respectively. To compute the functions $\tilde{v}_1$, $\tilde{w}_1$, $\tilde{v}_2$, and $\tilde{w}_2$ for $-\delta_0 \leq t \leq h - \varepsilon^\lambda$, $0 < \lambda < 1$, we can differentiate the asymptotic formulas in §§12.2–12.4 with respect to t and q, and then take the necessary linear combinations of them. For $t \geq h - \varepsilon^\lambda$ these functions must be found by integrating the system (13.1), (13.2), using the asymptotic formulas already found and the asymptotics of the relaxation cycle. In this way we get

$$a_{11}(\varepsilon) = \frac{A}{\beta}(1 - \exp(-\beta h)) + O(\varepsilon), \tag{13.8}$$

(13.9)
$$a_{12}(\varepsilon)\exp(-\delta_0) = \varepsilon\left\{\frac{A}{x_1}(1 - (1 + \alpha)x_1)\left(1 - \frac{1}{\beta}(1 - \exp(-\beta h))\right) + (A - A^2)\varkappa_1/x_1\right\} + O(\varepsilon^2),$$

$$a_{21}(\varepsilon)\exp\delta_0 = -\frac{A}{\alpha\beta\varepsilon}(1 - \exp(-\beta h))[1 + O(\varepsilon)], \tag{13.10}$$

$$a_{22}(\varepsilon) = A\left(1 + \frac{(1 - (1 + \alpha)x_1)}{\alpha\beta x_1}(1 - \exp(-\beta h))\right) - (A - A^2)\varkappa_1/\alpha x_1 + O(\varepsilon), \tag{13.11}$$

where, we recall,

$$\varkappa_1 = \frac{1}{\beta}(1 - \exp(-\beta h)) + x_1\exp(-\beta h).$$

We introduce the quadratic equation

$$\lambda^2 - \Delta_1\lambda + \Delta_2 = 0, \tag{13.12}$$

where

$$\Delta_1 = (1+\mu)A - A(1-A)\varkappa_1/\alpha x_1,$$
$$\Delta_2 = \mu A^2, \qquad \mu = \frac{1-x_1}{x_1}\frac{1}{\alpha\beta}[1-\exp(-\beta h)].$$

We remark that $\mu < 1$ by (11.30).

Below we assume that the roots $\lambda_1^0(A)$ and $\lambda_2^0(A)$ of equation (13.12) are simple. This restriction, which bears the character of generality of position, is not burdensome, since the complex parameter A takes finitely many values in what follows. We remark that the required property holds for $A = 1$, as follows from the formulas for Δ_1 and Δ_2 and the inequality $\mu < 1$.

The formulas (13.8)–(13.11) show that under the above simplicity restrictions for the roots of equation (13.12) the eigenvalues of the matrix (13.7) satisfy the asymptotic equalities

$$\lambda_s(\varepsilon) = \lambda_s^0(A) + O(\varepsilon), \qquad s = 1, 2.$$

All the facts we have presented are valid for the linear system associated with (13.1), (13.2). Therefore, exactly two multipliers of (13.1), (13.2) have the numbers $\lambda_1^0(A)$ and $\lambda_2^0(A)$ as their limits as $\varepsilon \to 0$, and the remaining multipliers approach zero exponentially. In correspondence with the end of §11.4 we now set $A = 1$, and this allows us to state the following assertion.

THEOREM 13.1. *Under the conditions of Theorem* 11.1 *the set* (11.15), (11.16) *contains exactly one initial condition of a relaxation cycle of the system* (11.1), (11.2) *that is orbitally exponentially stable.*

We discuss the question of stability of relaxation cycles with short periods, confining ourselves to the case $k = 1$. Under this condition equation (11.42) can be transformed to the form

$$\mu A^3 + (\mu - \varkappa_1/\alpha x_1)A^2 - A - 1 = 0. \tag{13.13}$$

It follows from (13.13) and the end of §11.4 that if the simplest relaxation cycle with short period loses stability as the parameters of the system (11.1), (11.2) change, then it happens in an oscillatory way. There thus arises the important question of the dynamics of (11.1), (11.2) in a neighborhood of the unstable relaxation cycle with short period. We give a heuristic solution of this problem.

As in §11.3, we work in the framework of zero approximations of the system (11.1), (11.2). Suppose that ξ, $\Delta\xi > 0$ and $\xi + \Delta\xi < h$. We assume that $y_0(t) \equiv 0$ for $-h < t < -\xi - \Delta\xi$ and $-\xi \le t \le 0$, respectively, and for $-\xi - \Delta\xi < t < -\xi$ we assume that

$$y_0(t) = \frac{1}{\alpha\beta} - \frac{1}{\alpha}\left(\frac{1}{\beta} - x_{\text{init}}\right)\exp[-\beta(t+\xi+\Delta\xi)]. \tag{13.14}$$

Here the parameter x_{init} is an arbitrary positive number. Suppose next that the value x_{in} is given at $t = 0$. Writing the formulas for $x_0(t)$, $y_0(t)$ exactly according to the rule in §11.3 for $t > 0$, we arrive at the four-dimensional mapping

$$\begin{aligned} x_{\mathrm{in}} &\to 1 - \exp[-\sigma(\Delta\xi, x_{\mathrm{init}})], \qquad & x_{\mathrm{init}} &\to x_{\mathrm{in}}, \\ \Delta\xi &\to h - \xi - \Delta\xi, \qquad & \xi &\to \sigma(\Delta\xi, x_{\mathrm{init}}), \end{aligned} \tag{13.15}$$

where the function σ is determined from the equation

$$\Phi(\Delta\xi, \sigma, x_{\mathrm{init}}) = 0. \tag{13.16}$$

Here, as in the proof of Lemma 11.2, we emphasize the dependence of the function (11.10) on the delay, whose role is played by $\Delta\xi$. A simple computation shows that the matrix of the mapping (13.15) linearized at the equilibrium state has one eigenvalue equal to zero, while the three others coincide with $1/A$, where the A are the solutions of (13.13).

Thus, the heuristically derived mapping (13.15) gives the same answer as the exact theory to the question of stability of the simplest relaxation cycle with short period.

We remark that the authors of [15] also posed the question of the existence and stability of relaxation cycles with short periods. Considering the simplest case and claiming rigor, but actually operating on a heuristic level, they introduced instead of the four-dimensional mapping (13.15) a certain three-dimensional mapping whose multipliers, computed at the equilibrium state, do not coincide with those supplied by the exact theory.

13.3. Concluding remarks. Thus, under the initial conditions (11.15), (11.16) the behavior of the system (11.1), (11.2) is in large respect analogous to that of a classical relaxation system, and the specific nature manifests itself in that the role of the relay element is played by the delayed feedback. At the same time, the system has a large number of relaxation cycles with short periods, and the technique for investigating them is fairly specific. Here a clarifying role would be played by a theorem on the rigorous connection between the mapping (13.15) and the solutions of the system (11.1), (11.2) with initial conditions of a certain form. Such a theorem would enable us, in particular, to make progress in the question of the existence of stable two-frequency relaxation oscillations for the system (11.1), (11.2), which the mapping (13.15) indicates are possible. We emphasize that progress in this direction can be achieved by an essential modernization of our technique. This has to do with the fact that above we constructed the asymptotics of relaxation oscillations under relatively simple initial conditions (the situation was rather different in the preceding chapter, and this led to many difficuties). There are thus still many open questions in the theory of relaxation oscillations for systems with delay.

14. Relaxation cycles of finite amplitude for a certain class of scalar equations with delay

14.1. Statement of the problem. The goal of this section is to illustrate the general methods developed in the investigation of the system (11.1), (11.2) for constructing the asymptotics and studying the stability of relaxation cycles, using as an example the equation

$$\varepsilon\dot{x} = xf(x, x(t-h)), \qquad 0 < \varepsilon \ll 1, \tag{14.1}$$

where $h > 0$, and the function $f(x, y) \in C^{\infty}(K)$, $K = \{(x, y) : x \geq 0, y \geq 0\}$, satisfies the following restrictions.

CONDITION 1. *There is an* $x_0 > 0$ *such that* $f(x, 0) > 0$ *for* $0 \le x < x_0$, $f(x, 0) < 0$ *for* $x > x_0$, *and* $f'_x(x_0, 0) < 0$.

CONDITION 2. $f'_y < 0$ *for* $x = x_0$ *and* $0 \le y \le y_0$.

CONDITION 3. $f(x, x_0) < 0$ *for* $0 \le x \le x_0$.

We remark that these conditions are satisfied for $\gamma > 1$ by the equation

(14.2) $$e\dot{x} = [1 - x - \gamma x(t-1)]x,$$

which is obtained from (2.21) after the substitution $\varepsilon = 1/\lambda$, $x = (a/(1+a))N$, $\gamma = 1/a$.

14.2. Existence of cycles. As in §11, we first pose the question of whether equation (14.1) has a slowly oscillating cycle $x(t, \varepsilon)$, that is, a cycle for which the equation $x(t, \varepsilon) = x_0/2$ has at most one solution on any time interval of length h. With this goal we distinguish in the phase space $C(-h, 0)$ a closed and bounded convex set S of initial conditions consisting of functions $\varphi(t, \varepsilon)$ such that

(14.3) $$0 \le \varphi(t, \varepsilon) \le (x_0/2)\exp(mt, \varepsilon), \qquad \varphi(0, \varepsilon) = x_0/2,$$

and the constant $m > 0$ is the same for all functions in the family.

Let $x_\varphi(t, \varepsilon)$ be a solution of (14.1) with arbitrary initial condition $\varphi \in S$. Under Conditions 1–3 the general asymptotic properties of $x_\varphi(t, \varepsilon)$ are easily revealed. Indeed, it follows from (14.3) that the solution $x_\varphi(t, \varepsilon)$ can be approximated up to quantities of order $\exp(-m\varepsilon^{\lambda-1})$ by the solution of the Cauchy problem

(14.4) $$\varepsilon\dot{x} = xf(x, 0), \qquad x|_{t=0} = x_0/2,$$

unformly with respect to $0 \le t \le h - \varepsilon^\lambda$, $0 < \lambda < 1$, and with respect to initial conditions $\varphi \in S$. Therefore (see Condition 1), the function $x_\varphi(t, \varepsilon)$ becomes asymptotically close to x_0 in a time of order $\varepsilon \ln(1/\varepsilon)$ and remains so up to values of t asymptotically close to h.

Conditions 1 and 2 ensure the inequalities $x_\varphi(t, \varepsilon) \le x_0$ for all $t \ge 0$. Indeed, if $\varphi(t, \varepsilon) > 0$ for $-h \le t \le 0$, then $x_\varphi(t, \varepsilon) < x_0$ for all $t > 0$. For if not, then there is a first moment of time $t_0 > 0$ such that $x_\varphi(t_0, \varepsilon) = x_0$ and $\dot{x}_\varphi(t_0, \varepsilon) \ge 0$. But, on the other hand, from (14.1) in view of Conditions 1 and 2 we have

$$\varepsilon\dot{x}_\varphi(t_0, \varepsilon) = x_0 f(x_0, x_\varphi(t_0 - h, \varepsilon)) < x_0 f(x_0, 0) = 0,$$

which contradicts the original premise. But if $\varphi(t, \varepsilon) \ge 0$, then clearly $x_\varphi(t, \varepsilon) \le x_0$.

Accordingly, the line $x = x_0$ forms a "barrier" for the solution $x_\varphi(t, \varepsilon)$. And this and Condition 3 give us that $x_\varphi(t, \varepsilon)$ begins to decrease for values of t a little larger than h, becomes a quantity of order $\exp\{(1/\varepsilon)f(0, x_0)(t-h)\}$ in a time of order $\varepsilon \ln(1/\varepsilon)$, and remains such up to an asymptotically small neighborhood of the point $t = 2h$.

For values of t just a little larger than $2h$ equation (14.4) again starts to operate, and $x_\varphi(t, \varepsilon)$ becomes a quantity of order $\exp\{(1/\varepsilon)f(0, x_0)h + (1/\varepsilon)f(0, 0)(t-2h)\}$. A complete compensation ensues in an asymptotically small neighborhood of $t = T_0$, where

(14.5) $$T_0 = [2 - f(0, x_0)/f(0, 0)]h > 2h;$$

then there is a "breakdown" onto the equilibrium state $x = x_0$ of equation (14.4), and so on.

Denote by $T(\varepsilon, \varphi)$ the second positive root of the equation $x_\varphi(t, \varepsilon) = x_0/2$. It follows from the above that, uniformly with respect to $\varphi \in S$,

$$T(\varepsilon, \varphi) = T_0 + o(1), \qquad \dot{x}_\varphi|_{t=T(\varepsilon, \varphi)} = (x_0/2\varepsilon)[f(x_0/2, 0) + o(1)]. \tag{14.6}$$

Therefore, the first-return operator

$$\Pi\varphi = x_\varphi(t + T(\varepsilon, \varphi), \varepsilon)$$

is defined on the set S, and it maps this set into itself for a suitable choice of $m > 0$ in view of (14.5), (14.6).

Theorem 14.1. *Under Conditions* 1–3 *there is an* $\varepsilon_0 > 0$ *such that for* $0 < \varepsilon < \varepsilon_0$ *equation* (14.1) *has a slowly oscillating relaxation cycle* $x(t, \varepsilon)$ *of period* $T(\varepsilon)$, $T(0) = T_0$.

To prove the existence of relaxation cycles $x_k(t, \varepsilon)$, $k = 1, 2, \dots,$ of equation (14.1) with periods $T_k(\varepsilon) < h/k$ we use the device in §11.4, which reduces to the following assertion in this case.

Theorem 14.2. *Assume Conditions* 1–3. *Then for any positive integer* k_0 *there is an* $\varepsilon_{k_0} > 0$ *such that for* $0 < \varepsilon < \varepsilon_{k_0}$ *equation* (14.1) *has rapidly oscillating relaxation cycles* $x_k(t, \varepsilon)$, $k = 1, \dots, k_0$, *whose periods* $T_k(\varepsilon)$ *satisfy the equalities*

$$T_k(\varepsilon) = (h - h_k)/k + o(1), \qquad h_k = h/(1 + kT_0).$$

14.3. Asymptotics of the cycle $x(t, \varepsilon)$ on the upward flight part and on the first slow motion part. In constructing the asymptotics of the cycle $x(t, \varepsilon)$ we single out the following parts: the upward flight part $-\delta_0 \le t \le \delta_0$, the first slow motion part $\delta_0 \le t \le h - \delta_0$, the drop part $h - \delta_0 \le t \le h + \delta_0$, and the second slow motion part $h + \delta_0 \le t \le T(\varepsilon) - \delta_0$. Here $\delta_0 > 0$ is sufficiently small and fixed.

As already mentioned, for $-\delta_0 \le t \le \delta_0$ we have the Cauchy problem (14.4), in which we make the substitution $t = (\varepsilon/\beta_0)\ln u$, where $\beta_0 = f(0, 0) > 0$. As a result we get the initial value problem

$$\beta_0 u v' = v f(v, 0), \qquad v|_{u=1} = x_0/2. \tag{14.7}$$

It follows from Conditions 1 and 2 that the function $v = v(u)$ is defined for all $0 < u < \infty$, is monotonically increasing, and satisfies $v(u) \to 0$ as $u \to 0$ and $v(u) \to x_0$ as $u \to \infty$. We dwell in greater detail on its asymptotic behavior as $u \to 0$ and as $u \to \infty$.

We remark that the problem (14.7) is equivalent to the equation

$$G(u) = \frac{1}{\beta_0}\ln u, \qquad G(v) = \int_{x_0/2}^{v} \frac{ds}{sf(s, 0)}, \tag{14.8}$$

and for small v the function G admits the representation

$$G(v) = \frac{1}{\beta_0}\ln v + g_0(v), \qquad g_0(v) = \int_{x_0/2}^{v} \frac{\beta_0 - f(s, 0)}{s\beta_0 f(s, 0)}\,ds - \frac{1}{\beta_0}\ln\frac{x_0}{2}. \tag{14.9}$$

Using (14.9) in (14.8) and setting $v = u\gamma(u)$, we get the equation

$$\frac{1}{\beta_0}\ln\gamma + g_0(u\gamma) = 0 \tag{14.10}$$

for determining γ. Applying the implicit function theorem to (14.10) for $u = 0$, we get a smooth function $\gamma = \gamma(u)$ with $\gamma(0) = \alpha_0$, where $\alpha_0 = \exp(-\beta_0 g_0(0))$.

As $u \to \infty$ we use the representation

$$
\begin{aligned}
G(v) &= -\frac{1}{\beta_1}\ln(x_0 - v) + g_1(v), \\
g_1(v) &= \frac{1}{\beta_1}\ln\frac{x_0}{2} + \int_{x_0/2}^{v} \frac{\beta_1(s - x_0) + sf(s, 0)}{\beta_1 s f(s, 0)(s - x_0)}\, ds, \\
\beta_1 &= -x_0 f'_x(x_0, 0) > 0.
\end{aligned} \tag{14.11}
$$

Using (14.11) in (14.8) and setting $u = 1/s$ and $v = x_0 - s^{\beta_1/\beta_0} z_0(s)$, we get the equation

$$-\frac{1}{\beta_1}\ln z_0 + g_1(x_0 - s^{\beta_1/\beta_0} z_0) = 0$$

for determining z_0, and from it we find with the help of the implicit function theorem that $z_0 = z_0(s) = \alpha + O(s^{\beta_1/\beta_0})$ as $s \to +0$, where $\alpha_1 = \exp(\beta_1 g_1(x_0))$. Thus, as $u \to \infty$,

$$v(u) = x_0 - \alpha_1/u^{\beta_1/\beta_0} + O(1/u^{2\beta_1/\beta_0}). \tag{14.12}$$

These constructions lead to the formulas

$$x(t, \varepsilon) = v(u), \qquad \exp(-\delta_0\beta_0/\varepsilon) \le u \le \exp(\delta_0\beta_0/\varepsilon), \tag{14.13}$$

$$x(t, \varepsilon) = x_0, \qquad \delta_0 \le t \le h - \delta_0. \tag{14.14}$$

The remainders here and in all the formulas below have order $\exp(-a/\varepsilon)$, $a > 0$, uniformly with respect to the indicated t, and hence we do not write them out.

14.4. Asymptotics of the cycle $x(t, \varepsilon)$ on the drop part. For $h - \delta_0 \le t \le h + \delta_0$ we make the substitutions $t = h + (\varepsilon/\beta_0)\ln u$, $w(u, \varepsilon) = x(t, \varepsilon)|_{t=h+(\varepsilon/\beta_0)\ln u}$ in (14.1), use the equality (14.13), and discard the exponentially small quantities. As a result we get

$$\beta_0 u w' = w f(w, v(u)), \tag{14.15}$$

which in view of (14.12) should be supplemented by the initial condition

$$w|_{u \to +0} = x_0. \tag{14.16}$$

Thus, the construction of the asymptotics of the cycle on the drop part has been reduced to the proof that the problem (14.15), (14.16) is solvable for $0 < u < \infty$, which is the main difficulty in the construction of the asymptotics as a whole.

To prove that the problem (14.15), (14.16) has a locally unique solution we let $w = x_0 + u\omega_0(u)$ and pass to the integral equation

$$
\begin{aligned}
L\omega_0(u) = \frac{1}{u}\int_0^u \bigg\{ &\frac{1}{\beta_0}\omega_0(s) f(x_0 + s\omega_0(s), v(s)) \\
&+ \frac{x_0}{\beta_0 s}[f(x_0 + s\omega_0(s), v(s)) - f'_x(x_0, 0)s\omega_0(s)] \bigg\}\, ds,
\end{aligned} \tag{14.17}
$$

where

$$L\omega_0(u) = \omega_0(u) + \frac{\beta_1}{\beta_0 u}\int_0^u \omega_0(s)\, ds. \tag{14.18}$$

If we then invert the operator (14.18) according to the formula

$$L^{-1}\omega_0(u) = \omega_0(u) - \frac{\beta_1}{\beta_0 u}\int_0^u \left(\frac{s}{u}\right)^{\beta_1/\beta_0} \omega_0(s)\,ds\,,$$

we get an integral equation whose right-hand side is denoted by $F(\omega_0)$. An uncomplicated analysis with the properties of the function $f(x, y)$ and the behavior of $v(u)$ for small u taken into account shows that the operator F transforms some subset of functions

$$\omega_0(u) \in C(0, u_0), \qquad \omega_0(0) = -\beta_1/(\beta_0 + \beta_1), \tag{14.19}$$

into itself, and is a contraction for sufficiently small u_0.

Thus, the problem (14.15), (14.16) is locally uniquely solvable, and $w(u) < x_0$ for all small $u > 0$ in view of (14.19). It follows from Conditions 1 and 2 (see the analogous place in §14.2) that $w(u) < x_0$ for all $u > 0$, and it follows from Condition 3 that for all sufficiently large u the function $w(u)$ is monotonically decreasing and satisfies

$$w(u) \sim c_0/u^{\beta_2}, \qquad \beta_2 = -f(0, x_0)/f(0, 0) > 0, \qquad u \to \infty. \tag{14.20}$$

Summarizing all the foregoing, we arrive at the formula

$$x(t, \varepsilon) = w(u), \qquad \exp(-\delta_0\beta_0/\varepsilon) \le u \le \exp(\delta_0\beta_0/\varepsilon), \tag{14.21}$$

where $t = h + (\varepsilon/\beta_0)\ln u$.

14.5. Asymptotics of the cycle $x(t, \varepsilon)$ on the second slow motion part. Since for $h + \delta_0 \le t \le T(\varepsilon) - \delta_0$ the function $x(t, \varepsilon)$ is exponentially small, it is more convenient to construct the asymptotics of $\ln x(t, \varepsilon)$ for these t. Using the asymptotic formulas already obtained and integrating (14.1) by steps, we get successively that

$$\ln x(t, \varepsilon) = f(0, x_0)(t - h)/\varepsilon + \ln c_0, \qquad h + \delta_0 \le t \le 2h - \delta_0, \tag{14.22}$$

where the constant c_0 is from (14.20), and

(14.23)
$$\ln x(t, \varepsilon) = f(0, x_0)\left(\frac{h}{\varepsilon} + \frac{1}{\beta_0}\ln u\right) + \int_0^u \frac{f(0, w(s)) - f(0, x_0)}{\beta_0 s}\,ds + \ln c_0,$$
$$t = 2h + \frac{\varepsilon}{\beta_0}\ln u, \qquad 2h - \delta_0 \le t \le 2h + \delta_0;$$

(14.24)
$$\ln x(t, \varepsilon) = f(0, x_0)\frac{h}{\varepsilon} + (H)\int_0^\infty \frac{f(0, w(s))}{\beta_0 s}\,ds + \ln c_0 + \frac{f(0, 0)}{\beta_0}\ln u,$$
$$t = 2h + \frac{\varepsilon}{\beta_0}\ln u, \qquad 2h + \delta_0 \le t \le T(\varepsilon) - \delta_0.$$

14.6. Computation of the period $T(\varepsilon)$. It follows from the general asymptotic properties of $x(t, \varepsilon)$ that for $T_0 - \delta_0 \le t \le T_0 + \delta_0$ we again get equation (14.7) after the substitution $t = T_0 + (\varepsilon/\beta_0)\ln u$. And from this and (14.24) we conclude that for such t

$$x(t, \varepsilon) = v(r_0 u/\alpha_0), \tag{14.25}$$

where

$$r_0 = \exp\left\{(H)\int_0^\infty \frac{f(0, w(s))}{\beta_0 s}\, ds + \ln c_0\right\}, \tag{14.26}$$

and v is the solution of the Cauchy problem (14.7). Finally, equating the right-hand side of (14.25) to $x_0/2$, we get

$$T(\varepsilon) = T_0 + \varepsilon T_1 + O(\exp(-a/\varepsilon)), \qquad a > 0, \tag{14.27}$$

where

$$T_1 = \frac{1}{\beta_0}(\ln\alpha_0 - \ln r_0). \tag{14.28}$$

In conclusion we note that it is unnecessary to construct the asymptotics of the cycles $x_k(t, \varepsilon)$, $k \geq 1$, because all the formulas in the preceding subsections are valid also for them. The only difference is that h is replaced by

$$\tilde{h} = h/[1 + k(T_0 + \varepsilon T_1)] + O(\exp(-a/\varepsilon)), \qquad a > 0,$$

and for their periods $T_k(\varepsilon)$ we have the formulas $T_k(\varepsilon) = (h - \tilde{h})/k$.

14.7. Investigation of stability of the relaxation cycles. It follows from §11.4 that the problem of stability of the cycles $x(t, \varepsilon)$ and $x_k(t, \varepsilon)$ reduces to an analysis of the distribution of the multipliers of the equation

$$\begin{aligned} \varepsilon\dot{v} &= x(t, \varepsilon)f'_x(x(t, \varepsilon), x(t-h, \varepsilon))v \\ &\quad + Af'_y(x(t, \varepsilon), x(t-h, \varepsilon))x(t-h, \varepsilon)v(t-h) \end{aligned} \tag{14.29}$$

with a complex parameter A. The investigation of (14.29) is carried out by the scheme in §13, and we do not dwell on it. We present the final result at once.

THEOREM 14.3. *For all sufficiently small ε equation* (14.29) *has a simple multiplier*

$$\nu(A, \varepsilon) = (1 - \varkappa)A + \varkappa A^2 + O(\exp(-a/\varepsilon)), \quad a > 0, \tag{14.30}$$

where $\varkappa = 1 - f(0, x_0)/f(0, 0) > 1$, and all the other multipliers are exponentially small.

Let us draw some conclusions. Setting $A = 1$ in (14.30) and using the fact that $\nu(1, \varepsilon) \equiv 1$, we conclude that the cycle $x(t, \varepsilon)$ of equation (14.1) is exponentially orbitally stable. Further, here the equation (11.42) responsible for stability of cycles with short periods takes the form

$$P(A) \equiv A^{1/k}[(1 - \varkappa)A + \varkappa A^2] = 1. \tag{14.31}$$

To analyze the distribution of its roots it is convenient to introduce the auxiliary equation

$$P(A, \lambda) = 1, \tag{14.32}$$

where $P(A, \lambda) = A^{1/k}[(1 - \lambda)A + \lambda A^2]$, $1 \leq \lambda \leq \varkappa$. We remark that, first, for all $1 \leq \lambda \leq \varkappa$ equation (14.32) has the simple root $A = 1$ ($\partial P/\partial A = 1/k + 1 + \lambda \neq 0$ for $A = 1$); second, for all $1 < \lambda \leq \varkappa$ equation (14.32) does not have other roots (besides $A = 1$) on the unit circle; finally, third, for $\lambda = 1$ the roots of (14.32) have the form

$$A_m = \exp(\pm i2\pi mk/(2k + 1)), \qquad m = 0, 1, \ldots, 2k + 1.$$

These properties of equation (14.32), together with the easily verified fact that for $0 < \lambda - 1 \ll 1$ all its roots (with the exception of $A = 1$, of course) shift inside the unit circle by quantities of order

$$\frac{2k}{2k+1}\left(1 - \cos\frac{2\pi km}{2k+1}\right)(\lambda - 1) > 0,$$

imply that they remain there also when $\lambda = \varkappa$.

We summarize all the foregoing as a theorem.

THEOREM 14.4. *Under Conditions* 1–3 *equation* (14.1) *has a unique exponentially orbitally stable slowly oscillating cycle* $x(t, \varepsilon)$ *for all sufficiently small* ε, *and all its rapidly oscillating cycles* $x_k(t, \varepsilon)$, $k \geq 1$, *are exponentially unstable.*

14.8. Stability of a homogeneous cycle in an environment with diffusion. The results obtained in §14.7 make it easy for us to solve the problem, important from an applied point of view, of diffusion instability of a cycle $x(t, \varepsilon)$ of equation (14.1). We divide (14.1) by ε, add the diffusion term $D\Delta x$ to its right-hand side, where $D > 0$, and pose Neumann boundary conditions on the sufficiently smooth boundary $\partial\Omega$ of a bounded domain $\Omega \subset R^m$, $m \geq 1$. As a result we get the parabolic boundary value problem

$$\frac{\partial x}{\partial t} = D\Delta x + \frac{x}{\varepsilon} f(x, x(t-h, s)), \qquad \left.\frac{\partial x}{\partial \nu}\right|_{\partial\Omega} = 0, \tag{14.33}$$

where $s \in R^m$, and ν is the direction of the outward normal to $\partial\Omega$.

By Theorem 14.4, if Conditions 1–3 hold, then the boundary value problem (14.33) has a spatially homogeneous (independent of s) cycle $x(t, \varepsilon)$ that is stable in the framework of the point model (14.1). The problem of its stability in the framework of the boundary value problem (14.33) reduces to an analysis of the distribution of the multipliers of the equation

$$\begin{aligned} \varepsilon\dot{v} = {} & x(t, \varepsilon) f'_x(x(t, \varepsilon), x(t-h, \varepsilon))v \\ & + f'_y(x(t, \varepsilon), x(t-h, \varepsilon))x(t-h, \varepsilon)v(t-h) - \varepsilon\sigma v, \end{aligned} \tag{14.34}$$

where $\sigma = \lambda_k D$, $k = 1, 2, \ldots$, and $0 < \lambda_1 \leq \lambda_2 \leq \ldots$ are the eigenvalues of the operator $-\Delta$ under the Neumann boundary conditions.

Assuming that the parameter $\sigma \geq 0$ in (14.34) varies continuously and making the substitution $v \exp(\sigma t) \to v$, we get equation (14.29), in which now $A = \exp(\sigma h)$. Therefore, all the multipliers of (14.34) tend to zero as $\varepsilon \to 0$ with the exception of one, which we denote by $\mu(\sigma, \varepsilon)$. From (14.30) we get for it the asymptotic equality

$$\mu(\sigma, \varepsilon) = [(1 - \varkappa)\exp(\sigma h) + \varkappa\exp(2\sigma h)]\exp(-\sigma T(\varepsilon)), \tag{14.35}$$

which is valid up to exponentially small quantities.

A simple analysis of the function (14.35) leads to the following assertion (assume that the phase space of the boundary value problem (14.33) is $C([-h, 0] \times \overline{\Omega})$, $\overline{\Omega} = \Omega + \partial\Omega$).

THEOREM 14.5. *Suppose that Conditions* 1–3 *hold and that the quantity* (14.28) *is negative. Then for all sufficiently small* ε *there is a*

$$D_{\text{cr}}(\varepsilon) = -\varepsilon T_1/\lambda_1\varkappa(\varkappa - 1)h^2 + O(\varepsilon^2)$$

such that a homogeneous cycle $x(t, \varepsilon)$ *of the boundary value problem* (14.33) *is exponentially orbitally stable (unstable) in the metric of the phase space* C *for* $D - D_{cr} > 0$ (< 0). *But if the quantity* (14.28) *is positive, the homogeneous cycle is stable for any values of the diffusion coefficient* D.

In conclusion we note that for the cycle $x(t, \varepsilon)$ of equation (14.2)

$$v(u) = u/(u+1), \qquad w(u) = (\gamma - 1)v(u)/[(u+1)^{\gamma-1} - 1],$$
$$T(\varepsilon) = \gamma + 1 - \varepsilon \ln(\gamma - 1).$$

Therefore, its diffusion instability is possible only when $\gamma > 2$ or, in biological variables, when $a < 1/2$.

15. The simplest relay with delay, and relaxation cycles of finite amplitude for a certain class of differential-difference equations

15.1. Statement of the problem. We recall once more that solutions of the relay system (6.13) serve as limits of solutions of the relaxation system of ordinary differential equations (6.12) in the C^1-metric as $\varepsilon \to 0$. In this section we establish precisely the same result for a certain class of differential-difference equations. However, a difference from the case of classical relaxation oscillations is that the limit object here is a relay with delay. The last circumstance confirms once more the idea put forth in the Introduction that it is impossible to construct a sufficiently general theory of relaxation oscillations for equations with delay, because of the great variety of limit objects and of asymptotic properties of solutions.

We consider the equation

$$\dot{x} = F(x(t-1)/\varepsilon), \qquad 0 < \varepsilon \ll 1, \tag{15.1}$$

where the function $F(x) \in C^{\infty}(-\infty, \infty)$ is such that

$$F(x) = 1 + \sum_{k=1}^{\infty} \frac{\alpha_k}{x^k}, \qquad F(x) = -\gamma_0 + \sum_{k=1}^{\infty} \frac{\beta_k}{x^k}, \qquad \gamma_0 > 0, \tag{15.2}$$

as $x \to -\infty$ and ∞, respectively. We assume further that the asymptotic series (15.2) can be differentiated any number of times.

Under these restrictions we pose the questions of existence of a limit object for equation (15.1) as $\varepsilon \to 0$ and of the character of convergence to it.

15.2. The C**-convergence theorem.** It follows from the asymptotic equalities (15.2) that

$$\lim_{\varepsilon \to 0} F(x/\varepsilon) = F_0(x), \qquad F_0(x) = \begin{cases} 1, & x < 0, \\ -\gamma_0, & x > 0. \end{cases}$$

Therefore, it is natural to expect that the relay equation

$$\dot{x} = F_0(x(t-1)) \tag{15.3}$$

is a limit object for equation (15.1).

Let $x_{\varphi}(t)$ be a solution of (15.3) with initial function φ in the set

$$S(r_1, r_2) = \{\varphi(t) \in C(-1, 0) : r_1(t+1) \le \varphi(t) \le r_2(t+1)\}, \tag{15.4}$$

where $0 < r_1 < 1$ and $r_2 > 1$, specified at $t = 0$. Then $x_\varphi(t) = -\gamma_0 t + \varphi(0)$ for $0 \le t \le t_1 + 1$, where $t_1 = \varphi(0)/\gamma_0$. As t passes through $t_1 + 1$ the function $x_\varphi(t-1)$ becomes negative ($\dot{x}_\varphi(t_1) = -\gamma_0 \neq 0$); that is, a switching takes place, and $x_\varphi(t) = x_\varphi(t_1+1) + t - t_1 - 1$ subsequently. A switching again takes place for $t = t_2 + 1$, where $t_2 = t_1 + 1 - x_\varphi(t_1+1)$, and so on.

Let $x_\varphi(t, \varepsilon)$ be a solution of (15.1) with initial function $\varphi(t) \in S(r_1, r_2)$.

THEOREM 15.1. *$x_\varphi(t, \varepsilon) \to x_\varphi(t)$ as $\varepsilon \to 0$, uniformly with respect to φ in the set* (15.4) *and t in any finite interval.*

For a proof we observe that (15.1) implies the estimate

$$|x_\varphi(t, \varepsilon) - \varphi(0)| \le M\varepsilon^\lambda, \qquad M = \sup|F(x)|, \ -\infty < x < \infty,$$

for $0 \le t \le \varepsilon^\lambda$, $0 < \lambda < 1$, and it follows from the conditions (15.2) that, up to quantities of order,*

$$x_\varphi(t, \varepsilon) = x_\varphi(\varepsilon^\lambda, \varepsilon) - \gamma_0(t - \varepsilon^\lambda), \qquad \varepsilon^\lambda \le t \le 1 + x_\varphi(\varepsilon^\lambda, \varepsilon)/\gamma_0,$$

uniformly with respect to $\varphi \in S(r_1, r_2)$ and the indicated t. Continuing the process, we get the theorem.

We point out that the solution $x_0(t)$ of (15.3) with the initial function $\varphi = t + 1$ (which belongs to $S(r_1, r_2)$ by the choice of r_1 and r_2) generates a cycle of period $T_0 = 2 + \gamma_0 + 1/\gamma_0$ to which all the trajectories $x_\varphi(t)$ with $\varphi \in S(r_1, r_2)$ "attach" in a finite time. This and Theorem 15.1 yield the following assertion.

THEOREM 15.2. *There is an $\varepsilon_0 > 0$ such that for all $0 < \varepsilon < \varepsilon_0$ equation* (15.1) *has a cycle $x_0(t, \varepsilon)$, $x_0(-1, \varepsilon) = 0$, of period $T(\varepsilon)$, and*

$$\lim_{\varepsilon \to 0} T(\varepsilon) = T_0, \qquad \lim_{\varepsilon \to 0} \max_t |x_0(t, \varepsilon) - x_0(t)| = 0.$$

Thus, equation (15.3) does serve as a limit object for (15.1), and Theorem 15.1 gives a precise meaning to this concept.

15.3. Asymptotics of the cycle $x_0(t, \varepsilon)$. To construct the asymptotics of the cycle $x_0(t, \varepsilon)$ we note that $x_0(t-1) < 0$ for $-1 - \delta_0 \le t \le -\delta_0$, where $0 < 2\delta_0 < \min(1, \gamma_0)$. Taking into account this circumstance (together with the properties of $F(x)$ as $x \to -\infty$) in equation (15.1), we see that

$$(15.5) \qquad x_0(t, \varepsilon) = t + 1 + \varepsilon\Delta_1(t, \varepsilon), \qquad \Delta_1(-1, \varepsilon) \equiv 0; \qquad |\Delta_1| + |\dot{\Delta}_1| \le M,$$

for $-1 - \delta_0 \le t \le -\delta_0$, where the same letter M is used here and below to denote different universal constants.

Integrating (15.1) by steps with the initial condition (15.5) and again using the property (15.2), we get successively that for $-\delta_0 \le t \le \delta_0$

$$(15.6) \qquad x_0(t, \varepsilon) = 1 - \alpha_1 \varepsilon \ln(1/\varepsilon) + \varepsilon v(\tau) + \varepsilon^2 \omega_1(\tau, \varepsilon), \qquad t/\varepsilon = \tau,$$

where

$$v(\tau) = \int_0^\tau F(s)\,ds, \qquad |\omega_1| \le M \ln(|\tau| + 2), \qquad |\dot{\omega}_1| \le M/(|\tau| + 2);$$

* *Editor's note.* There is an omission at this point in the Russian original.

for $\delta_0 \le t \le t_1(\varepsilon) + 1 - \delta_0$

(15.7) $x_0(t, \varepsilon) = -\gamma_0 t + 1 + (\beta_1 - \alpha_1)\varepsilon \ln(1/\varepsilon) + \varepsilon\Delta_2(t, \varepsilon), \qquad |\Delta_2| + |\dot{\Delta}_2| \le M,$

where $t_1(\varepsilon) = 1/\gamma_0 + ((\beta_1 - \alpha_1)/\gamma_0)\varepsilon \ln(1/\varepsilon) + O(\varepsilon)$ is the first positive zero of $x_0(t, \varepsilon)$; for $t_1(\varepsilon) + 1 - \delta_0 \le t \le t_1(\varepsilon) + 1 + \delta_0$

(15.8) $$x_0(t, \varepsilon) = -\gamma_0 + [(1 + 1/\gamma_0)\beta_1 - \alpha_1]\varepsilon \ln(1/\varepsilon) + \varepsilon w(\tau) + \varepsilon^2 \omega_2(\tau, \varepsilon),$$

where $\tau = (t - t_1(\varepsilon) - 1)/\varepsilon$, and

$$w(\tau) = \int_0^\tau F(-\gamma_0 s)\, ds, \qquad |\omega_2| \le M \ln(|\tau| + 2), \qquad |\dot{\omega}_2| \le M/(|\tau| + 2);$$

for $t_1(\varepsilon) + 1 + \delta_0 \le t \le t_2(\varepsilon) + 1 - \delta_0$

(15.9) $$x_0(t, \varepsilon) = t - 1 - \gamma_0 - 1/\gamma_0 + (\beta_1 - \alpha_1)\varepsilon \ln(1/\varepsilon) + \varepsilon\Delta_3(t, \varepsilon),$$

where $t_2(\varepsilon)$ is the second positive zero of $x_0(t, \varepsilon)$ and $|\Delta_3| + |\dot{\Delta}_3| \le M$; and, finally,

(15.10) $$T(\varepsilon) = t_2(\varepsilon) + 1 = 2 + \gamma_0 + 1/\gamma_0 + (\alpha_1 - \beta_1)\varepsilon \ln(1/\varepsilon) + O(\varepsilon).$$

15.4. Stability properties of the cycle $x_0(t, \varepsilon)$. Let $h(t, \varepsilon)$ be a solution of the variational equation

(15.11) $$\dot{h} = (1/\varepsilon)F'(x_0(t - 1, \varepsilon)/\varepsilon)h(t - 1)$$

with arbitrary initial function $h_0(t)$ in the ball of unit radius about zero in the phase space $C(-1 - \delta_0, -\delta_0)$, where δ_0 is from (15.5). Integrating equation (15.11) by steps and using the asymptotics (15.5)–(15.9) and the properties (15.2), we arrive at the asymptotic representation

(15.12) $$h(t, \varepsilon) = h_0(-\delta_0)\dot{x}_0(t - 1, \varepsilon) + O(\varepsilon),$$

which is valid uniformly with respect to $-\delta_0 \le t \le T(\varepsilon) - \delta_0$ and with respect to the initial functions $h_0(t)$. The solutions of the equation associated with (15.11) have an analogous structure. Therefore, the unit multiplier of equation (15.11) is simple, and all its other multipliers lie inside a disk with radius of order ε.

THEOREM 15.3. *The cycle supplied by Theorem* 15.2 *for equation* (15.1) *is unique and exponentially orbitally stable.*

Let $0 = \tau_0 < \tau_1 < \cdots$ be the successive switching times of the cycle $x_0(t)$ for the relay equation (15.3). The formula (15.12) shows that, uniformly with respect to t outside certain sufficiently small fixed neighborhoods of the points τ_k ($k \ge 0$) the function $h(t, \varepsilon)$ converges to the solution $h(t)$ of the Cauchy problem

(15.13) $$\dot{h} = 0, \qquad h(\tau_k + 0) = \frac{\dot{x}_0(\tau_k + 0)}{\dot{x}_0(\tau_k - 0)} h(\tau_k - 0), \quad k \ge 0, \qquad h|_{t=-\delta_0} = h_0(-\delta_0).$$

And since (15.13) is the linearization of the relay equation (15.3) on the cycle $x_0(t)$, it follows that C^1-closeness as $\varepsilon \to 0$ has been established for periodic solutions of (15.1) and (15.3). It can be shown that C^1-closeness, in the sense indicated above (cf. [28]), holds for any trajectories of (15.1) and (15.3) with initial conditions (15.4).

15.5. On an equation in biophysics. As an example we consider the equation

$$(15.14)\qquad \varepsilon\dot{u} = [-1 + \alpha f(u(t-1)) - \beta g(u)]u, \qquad \alpha, \beta > 0, \ \alpha > 1 + \beta,$$

where $0 < \varepsilon \ll 1$, which arises in modelling the functioning of an individual neuron [55]. Here the positive functions $f(u)$, $g(u) \in C^2[0, \infty)$ are such that $f(0) = g(0) = 1$, $\beta g(u) + 1 < \alpha$ for all $u \geq 0$, and

$$f, g, uf', ug', u^2 f'', u^2 g'' = O(1/u), \qquad u \to \infty.$$

Making the substitution $u = \exp(x/\varepsilon)$ in (15.14), we get the relay equation

$$(15.15)\qquad \dot{x} = -1 + \alpha R_0(x(t-1)) - \beta R_0(x), \qquad R_0(x) = \begin{cases} 0, & x > 0, \\ 1, & x < 0, \end{cases}$$

as $\varepsilon \to 0$; its solution is defined similarly to that in §15.2. It is easy to verify that, like (15.3), equation (15.15) has a stable cycle

$$x_0(t) = (\alpha - 1)(t + 1), \quad -1 \leq t \leq 0, \qquad x_0(t) = \alpha - 1 - t, \quad 0 \leq t \leq \alpha - 1,$$
$$x_0(t) = -(1 + \beta)(t - \alpha + 1), \quad \alpha - 1 \leq t \leq \alpha,$$
$$x_0(t) = -1 - \beta + (\alpha - \beta - 1)(t - \alpha), \quad \alpha \leq t \leq \alpha + (1 + \beta)/(\alpha - \beta - 1)$$

of period $T_0 = \alpha + 1 + (1 + \beta)/(\alpha - \beta - 1)$. Therefore, there are analogues of Theorems 15.1–15.3, but here the moduli of the multipliers of the equation analogous to (15.11) (with the exception of the simple unit) have orders $\exp(-a/\varepsilon)$, $a > 0$.

In conclusion we note that the technique developed in this chapter enables us also to solve the problem of diffusion instability of a relaxation cycle of equation (15.14), which is important from the viewpoint of applications. Omitting the corresponding computations, we mention only that the answer here will be the same as in §14, that is, diffusion instability appears or does not appear depending on the sign of T_1, where T_1 is a correction of order ε to the period of a cycle.

Conclusion

First we touch once again on the history of the problem. Individual results on the existence of relaxation cycles in systems with delay have been known previously, of course. Our contribution to the C-theory is that we show, using difficult examples in mathematical ecology, the important role not only of the existence of a limit object, but also of the topology in which the convergence takes place. As for the C^1-theory, we do not know of any publications containing a sufficiently rigorous investigation of the variational equations. In this connection we recall once more that even for systems of ordinary equations this problem, posed long ago by Pontryagin and Mishchenko, was solved only quite recently.

We note especially the results in Chapter II, which have a fundamental character: it was shown for the first time that all the multipliers of the variational equation can stand on equal footing; that is, a phenomenon was established for which there is no immediate analogue in ordinary equations. Also, most of the problems we posed are connected with the results in that chapter. We mention some questions that have not been investigated.

For example, it is of essential interest to analyze the properties of relaxation cycles of the system

$$\dot{N}_1 = \frac{\lambda}{1+a}[1 + a(1 - N_2) - N_1(t-1)]N_1,$$
$$\dot{N}_2 = \frac{\lambda}{1+a}[1 + a(1 - N_2) - N_1(t-1)]N_2,$$

λ a large parameter, which models a competitive struggle between two strongly fertile twin species.

One more interesting problem is the analogous question for the system

$$\dot{N}_1 = \frac{\lambda}{1+a_1+a_2}[1 + a_1(1 - N_2) + a_2(1 - N_3) - N_1(t-1)]N_1,$$
$$\dot{N}_2 = r_2[N_1 + a_{23}(N_1 - N_3) - N_2(t - h_2)]N_2,$$
$$\dot{N}_3 = r_3[N_1 + a_{32}(N_1 - N_3) - N_3(t - h_3)]N_3,$$

λ again a large parameter, which models the interaction of two competing predators with their strongly fertile prey.

We regard it as impossible to construct a general theory of relaxation oscillations in systems with delay. Moreover, the construction of such a theory seems only to be possible for systems having a certain nonformal inner logic, as is nicely seen from the example of ecological equations. It would be interesting to discover new classes of differential-difference equations, far removed from ecological equations, for which a theory of relaxation oscillations can be created.

We turn again to ecological equations. As for their general characteristics, we must point out that they are fundamentally nonlocal. The latter means that in most cases approaches involving the use of any familiar asymptotic methods do not give sufficiently complete information about the character and structure of their attractors. As a typical example we cite the predator-prey problem investigated in §5, where the results obtained with the help of one of the most "nonlinear" asymptotic methods—the theory of relaxation oscillations—lead only to a determination of the limit behavior of the solutions, while the main events take place for moderate values of the Malthusian coefficient of linear growth of the prey. We encounter a quite different situation in Chapter II, where, vice versa, the asymptotic theory works nicely for moderate values of the auxiliary energy parameter, but becomes unsuitable as it approaches 1. It is remarkable that this happens even in the case when we are theoretically dealing with strongly unstable relaxation cycles that, however, are stable for intermediate values of the energy parameter. In extending our results to other ecological problems (for example, to those cited above) one should thus be careful; in particular, one should certainly make wide use of numerical experiments.

Bibliography

1. G. Ya. Beĭ-Bienko, *General entomology*, "Vysshaya Shkola", Moscow, 1980. (Russian)
2. B. P. Belousov, *A periodically acting reaction and its mechanism*, Auto-wave Processes in Systems with Diffusion (M. T. Grekkova, ed.), Inst. Prikl. Fiz. Akad. Nauk SSSR, Gorki, 1981, pp. 176–186. (Russian)
3. A. T. Bergerud, *Prey switching in a simple ecosystem*, Sci. Amer. **249** (1983), no. 6, 130–141.
4. V. O. Boldoruev, *Dynamics of the number of Siberian silkworms and their parasites*, Buryat. Knig. Izdat., Ulan-Udè, 1969. (Russian)
5. G. C. Varley, G. R. Gradwell, and M. P. Hassell, *Insect population ecology: an analytical approach*, Blackwell Sci. Publ., Oxford, and Univ. of California Press, Berkeley, CA, 1974.
6. G. A. Viktorov, *Problems in the dynamics of the number of insects, using the example of Eurygaster Integriceps*, "Nauka", Moscow, 1976. (Russian)
7. ———, *Ecology of parasites-insectivores*, "Nauka", Moscow, 1976. (Russian)
8. Vito Volterra, *Leçons sur la théorie mathématique de la lutte pour la vie*, Gauthier-Villars, Paris, 1931.
9. A. I. Vorontsov, *Forest entomology*, 3rd ed., "Vysshaya Shkola", Moscow, 1975. (Russian)
10. F. R. Gantmakher, *Sur les noyaux de Kellogg non symétriques*, C. R. (Dokl.) Acad. Sci. URSS **1 (10)** (1936), 3–5.
11. S. C. Glyzin, *Two-frequency oscillations of the fundamental equation for the dynamics of insect populations*, Nonlinear Oscillations and Ecology (Yu. S. Kolesov, ed.), Yaroslav. Gos. Univ., Yaroslavl, 1984, pp. 91–116. (Russian)
12. ———, *Stationary regimes of a finite-difference approximation of the Hutchinson equation with diffusion*, Qualitative and Approximate Methods for Studying Operator Equations (V. S. Klimov, ed.), Yaroslav. Gos. Univ., Yaroslavl, 1986, pp. 112–127. (Russian)
13. S. D. Glyzin and Yu. S. Kolesov, *An attractor of a biolocal model of the Hutchinson equation with diffusion when the coefficient of linear growth is large*, Dokl. Akad. Nauk SSSR **307** (1989), 351–353; English transl. in Soviet Phys. Dokl. **34** (1989).
14. ———, *An optimal method for fishery management*, Mathematical Models in Biology and Medicine (A. Yu. Garlyauskas [Argis Garlauskas] and D. I. Shivtra [Donatas Šivtra], eds.), no. 3, Inst. Mat. i Kibernet. Akad. Nauk Litovsk. SSR, Vilnius, 1989, pp. 37–41. (Russian)
15. E. V. Grigor′eva et al., *Application of asymptotic methods to the study of steady-state regimes of generation in lasers with a nonlinear delay element*, Mat. Model. **2** (1990), no. 4, 97–120. (Russian)
16. Rimantas Grikenis [Grikienis], *Modeling an ecosystem of insects with superparasitism taken into account*, Mathematical Models in Biology and Medicine (A. Yu. Garlyauskas [Argis Garlauskas] and D. I. Shivtra [Donatas Šivtra], eds.), no. 3, Inst. Mat. i Kibernet. Akad. Nauk Litovsk. SSR, Vilnius, 1989, pp. 42–56. (Russian)
17. A. V. Egorov, *Hoofed wildlife of Yakutia*, "Nauka", Moscow, 1974. (Russian)
18. A. M. Zhabotinskiĭ, *Concentration oscillations*, "Nauka", Moscow, 1968. (Russian)
19. K. M. Zavadskiĭ, *Species and the formation of species*, "Nauka", Moscow, 1968. (Russian)
20. S. A. Kashchenko, *Stationary regimes in the predator-prey problem*, Preprint No. 54–84, Inst. Mat. Akad. Nauk Ukrain. SSR, Kiev, 1984. (Russian) R. Zh. Mat. **1985**, 5Б375
21. A. Yu. Kolesov, *The domain principle in the problem of fluctuations in a mammal population*, Nonlinear Oscillations in Problems in Ecology (Yu. S. Kolesov, ed.), Yaroslav. Gos. Univ., Yaroslavl, 1985, pp. 11–22. (Russian)
22. ———, *On stability of a spatially homogeneous cycle of the Hutchinson equation with diffusion*, Mathematical Models in Biology and Medicine, no. 1, Inst. Mat. i Kibernet. Akad. Nauk Litovsk. SSR, Vilnius, 1985, pp. 93–102. (Russian)

23. ———, *A criterion for the stability of traveling waves of parabolic systems with small diffusion*, Sibirsk. Mat. Zh. **30** (1989), no. 3, 175–179; English transl. in Siberian Math. J. **30** (1989).

24. ———, *On instability of duck-cycles arising when an equilibrium state of a multidimensional relaxed system passes through a failure manifold*, Uspekhi Mat. Nauk **44** (1989), no. 5 (269), 165–166; English transl. in Russian Math. Surveys **44** (1989).

25. A. Yu. Kolesov and Yu. S. Kolesov, *Relaxation cycles of differential-difference equations*, Izv. Ross. Akad. Nauk Ser. Mat. **56** (1992), 790–812; English transl. in Russian Acad. Sci. Izv. Math. **41** (1993).

26. ———, *Relaxation cycles in systems with delay*, Mat. Sb. **183** (1992), no. 8, 141–159; English transl. in Math. USSR Sb. **76** (1993).

27. A. Yu. Kolesov, Yu. S. Kolesov, and V. V. Maĭorov, *The Belousov reaction: a mathematical model and experimental facts*, Dynamics of Biological Populations, Gor′kov. Gos. Univ., Gorki, 1987, pp. 43–51. (Russian)

28. A. Yu. Kolesov et al., *Asymptotic integration of the variational system of a multidimensional relaxation cycle*. I, II, Differentsial′nye Uravneniya **23** (1987), 1881–1889, 2036–2047; English transl. in Differential Equations **23** (1987).

29. ———, *A relaxation system in a neighborhood of a point of discontinuity: reduction to the regular case*, Uspekhi Mat. Nauk **43** (1988), no. 2 (260), 141–142; English transl. in Russian Math. Surveys **43** (1988).

30. ———, *On chaos phenomena in three-dimensional relaxation systems*, Mat. Zametki **46**, no. 2, 153–155. (Russian)*

31. A. Yu. Kolesov and E. F. Mishchenko, *Asymptotics of relaxation oscillations*, Mat. Sb. **137 (179)** (1988), 3–18; English transl. in Math. USSR Sb. **65** (1990).

32. Yu. S. Kolesov, *A survey of results in the theory of stability of solutions of differential-difference equations with almost periodic coefficients*, Studies in Stability and the Theory of Oscillations (Yu. S. Kolesov and P. P. Zabreĭko, eds.), Yaroslav. Gos. Univ., Yaroslavl, 1977, pp. 82–141. (Russian)

33. ———, *Mathematical models in ecology*, Studies in Stability and the Theory of Oscillations (Yu. S. Kolesov, ed.), Yaroslav. Gos. Univ., Yaroslavl, 1979, pp. 3–40. (Russian)

34. ———, *On stability of solutions of linear differential-difference equations of neutral type*, Sibirsk. Mat. Zh. **20** (1979), 317–321; English transl. in Siberian Math. J. **20** (1979).

35. ———, *Asymptotics of a periodic solution of the Hutchinson equation*, Factors of Diversity in Mathematical Ecology and Population Genetics (F. S. Berezovskaya, ed.), Nauchno-Issled. Vychisl. Tsentr. Akad. Nauk SSSR, Pushchino, 1980, pp. 54–62. (Russian)

36. ———, *The influence of hunting on the dynamics of the variation in the number of a species*, Studies in Stability and the Theory of Oscillations (Yu. S. Kolesov, ed.), Yaroslav. Gos. Univ., Yaroslavl, 1981, pp. 129–130. (Russian)

37. ———, *On some problems in the theory of oscillations*, Studies in Stability and the Theory of Oscillations, Yaroslav. Gos. Univ., Yaroslavl, 1982, pp. 20–41. (Russian)

38. ———, *Properties of solutions of a class of equations with delay that describe the dynamics of variation in the number of a species, with the age structure taken into account*, Mat. Sb. **117 (159)** (1982), 86–94; English transl. in Math. USSR Sb. **45** (1983).

39. ———, *Modeling of insect populations*, Biofizika **28** (1983), 513–514; English transl. in Biophysics **28** (1983).

40. ———, *The parasite-host problem*, Dynamics of Biological Populations, Gor′kov. Gos. Univ., Gorki, 1984, pp. 16–29. (Russian)

41. ———, *Stability and bifurcation of traveling waves*, Nonlinear Oscillations in Problems of Ecology (Yu. S. Kolesov, ed.), Yaroslav. Gos. Univ., Yaroslavl, 1985, pp. 3–10. (Russian)

42. ———, *The fundamental constants determining the dynamics of populations in a heterogeneous environment*, Dynamics of Biological Populations, Gor′kov. Gos. Univ., Gorki, 1988, pp. 105–106. (Russian)

43. ———, *Null asymptotics of the relaxation cycle of a mathematical model of the Belousov reaction*, Modeling of Population Dynamics (D. V. Goryachenko, ed.), Nizhegorod. Gos. Univ., Nizhnii Novgorod, 1990, pp. 28–35. (Russian)

44. ———, *Analysis of a mathematical model of ecology*, Dokl. Akad. Nauk SSSR **316** (1991), 577–580; English transl. in Soviet Phys. Dokl. **36** (1991).

45. ———, *Relaxation cycles of a differential-difference equation in the dynamics of insect populations*, Mathematical Models in Biology and Medicine, no. 4, Inst. Mat. i Kibernet. Akad. Nauk Litovsk. SSR, Vilnius, 1991, pp. 26–40. (Russian)

Editor's note.* According to MR **91a:58162, the content of this note is the same as their paper, *A mixing attractor in relaxation systems*, Dokl. Akad. Nauk SSSR **306** (1989), 38–40; English transl. in Soviet Math. Dokl. **39** (1989).

46. Yu. S. Kolesov and A. Yu. Kolesov, *Biological and physical parabolic systems*, Mathematical Models in Biology and Medicine, no. 2 (D. I. Shvitra [Donatas Švitra], ed.), Inst. Mat. i Kibernet. Akad. Nauk Litovsk. SSR, Vilnius, 1987, pp. 5–25. (Russian)
47. Yu. S. Kolesov, V. S. Kolesov, and I. I. Fedik, *Auto-oscillations in systems with distributed parameters*, "Naukova Dumka", Kiev, 1979. (Russian)
48. Yu. S. Kolesov and V. V. Maĭorov, *The role of an ecological trap in stabilization of the biological community of Newfoundland*, Nonlinear Oscillations in Problems of Ecology (Yu. S. Kolesov, ed.), Yaroslav. Gos. Univ., Yaroslavl, 1985, pp. 93–102. (Russian)
49. ———, *Spatial and temporal self-organization in a single-species biocenosis*, Dynamics of Biological Populations, Gor′kov. Gos. Univ., Gorki, 1986, pp. 3-13. (Russian)
50. Yu. S. Kolesov and D. I. Shvitra [Donatas Švitra], *Auto-oscillations in systems with delay*, "Mokslas", Vilnius, 1979. (Russian)
51. N. G. Kolomiets, A. I. Vorontsov, and G. V. Stadnitskiĭ, *The fox-colored pine sawfly*, "Nauka", Novosibirsk, 1972; English transl., *The European pine sawfly*, Amerind Publ. Co., New Delhi, and U.S. Dept. Agriculture and National Tech. Inform. Service, Springfield, VA, 1979.
52. Yu. P. Kondakov, *Regularities of mass reproduction of the Siberian silkworm*, The Ecology of Populations of Forest Animals of Siberia (E. S. Petrenko, ed.), "Nauka", Novosibirsk, 1974, pp. 206–265. (Russian)
53. E. P. Kubyshkin, *On the problem of the existence of single-species ichthyocenoses*, Studies in Stability and the Theory of Oscillations (Yu. S. Kolesov, ed.), Yaroslav. Gos. Univ., Yaroslavl, 1983, pp. 90–93. (Russian)
54. David Lambert Lack, *The natural regulation of animal numbers*, Clarendon Press, Oxford, 1954.
55. V. V. Maĭorov and I. Yu. Myshkin, *On a model of the functioning of a neuron network*, Modeling of Population Dynamics (V. D. Goryachenko, ed.), Nizhegorod. Gos. Univ., Nizhnii Novgorod, 1990, pp. 70–78. (Russian)
56. Ernst Mayr, *Animal species and evolution*, Harvard Univ. Press, Cambridge, MA, 1963.
57. E. N. Matyushkin, *The lynx*, "Lesnaya Promyshlennost′", Moscow, 1974. (Russian)
58. E. F. Mishchenko, *Asymptotic computation of periodic solutions of systems of differential equations containing small parameters in the derivatives*, Izv. Akad. Nauk SSSR Ser. Mat. **21** (1957), 627–654; English transl. in Amer. Math. Soc. Transl. (2) **18** (1961).
59. E. F. Mishchenko and L. S. Pontryagin, *Periodic solutions of systems of differential equations close to discontinuous solutions*, Dokl. Akad. Nauk SSSR **102** (1955), 889–891. (Russian)
60. ———, *Proof of some asymptotic formulas for solutions of differential equations with a small parameter*, Dokl. Akad. Nauk SSSR **120** (1958), 967–969. (Russian)
61. ———, *Derivation of some asymptotic estimates for solutions of differential equations with a small parameter in the derivatives*, Izv. Akad. Nauk SSSR Ser. Mat. **23** (1959), 643–660. (Russian)
62. E. F. Mishchenko and N. Kh. Rozov, *Differential equations with small parameters, and relaxation oscillations*, "Nauka", Moscow, 1975; English transl., Plenum Press, New York, 1980.
63. L. S. Pontryagin, *The asymptotic behavior of solutions of systems of differential equations with a small parameter in the highest derivatives*, Izv. Akad. Nauk SSSR Ser. Mat. **21** (1957), 605–626; English transl. in Amer. Math. Soc. Transl. (2) **18** (1961).
64. B. N. Sadovskiĭ, *Limit-compact and condensing operators*, Uspekhi Mat. Nauk **27** (1972), no. 1 (163), 81–146; English transl. in Russian Math. Surveys **27** (1972).
65. Yu. M. Svirezhev and N. V. Timofeev-Resovskiĭ, *Opposing selection pressures on genotypes and characters in sex-linked mutations*, Problemy Kibernet. Vyp. **18** (1967), 155–170; English transl. in Systems Theory Res. **18** (1968).
66. V. A. Tavrovskiĭ (ed.), *Mammals of Yakutia*, "Nauka", Moscow, 1971. (Russian)
67. R. S. Ushatinskaya and G. G. Ĭirkovskiĭ, *Ecology and physiology of the Colorado* [*potato*] *beetle*, "Nauka", Moscow, 1976. (Russian)
68. N. B. Fedotov, *Peculiarities of fluctuations of the squirrel population and the population of their main food competitors*, Studies in Stability and the Theory of Oscillations (Yu. S. Kolesov, ed.), Yaroslav. Gos. Univ., Yaroslavl, 1983, pp. 44–54. (Russian)
69. A. N. Sharkovskiĭ, Yu. L. Maĭstrenko, and E. Yu. Romanenko, *Difference equations and their applications*, "Naukova Dumka", Kiev, 1986; rev. English transl., Kluwer, Dordrecht, 1993.
70. S. S. Shvarts, *Ecological regularity of evolution*, "Nauka", Moscow, 1980. (Russian)
71. V. I. Yudovich, *Spectral properties of the evolution operator of a parabolic equation with one spatial variable and of one-dimensional analogues of it*, Proc. All-Union Conf. Partial Differential Equations (on the 75th Anniversary of the Birth of I. G. Petrovskiĭ), Mekh.-Mat. Fak. Moskov. Gos. Univ., Moscow, 1978, pp. 497–498. (Russian)

72. A. V. Yablokov, *Phenetics: evolution, population, trait*, "Nauka", Moscow, 1980; English transl., Columbia Univ. Press, New York, 1986.
73. Jack Hale, *Theory of functional differential equations*, 2nd ed, Springer-Verlag, Berlin, 1977.
74. G. Hoffman de Visme, *The density of prime numbers*, Math. Gaz. **45** (1961), 13–14.
75. G. Evelyn Hutchinson, *Circular causal systems in ecology*, Ann. New York Acad. Sci **50** (1948–1950), 221–246.
76. G. Stephen Jones, *Asymptotic behavior and periodic solutions of a non-linear differential-difference equation*, Proc. Nat. Acad. Sci. U.S.A. **47** (1961), 879–882.
77. O. D. Kellogg, *The oscillation of functions on an orthogonal set*, Amer. J. Math. **38** (1916), 1–5.
78. ———, *Orthogonal function sets arising from integral equations*, Amer. J. Math. **40** (1918), 145–154.
79. Lloyd B. Keith, *Wildlife's ten-year cycle*, Univ. of Wisconsin Press, Madison, WI, 1963.
80. E. M. Wright, *A non-linear differential-difference equation*, J. Reine Angew. Math. **194** (1955), 66–87.
81. ———, *A functional equation in the heuristic theory of primes*, Math. Gaz. **44** (1960), 15–16.